Collins

Student Support Materials for AQA

A2 Chemistry

Unit 5: Energetics, Redox and Inorganic Chemistry

Authors: John Bentham, Colin Chambers, Graham Curtis,

Geoffrey Hallas, Andrew Maczek, David Nichol

William Collins's dream of knowledge for all began with the publication of his first book in 1819. A self-educated mill worker, he not only enriched millions of lives, but also founded a flourishing publishing house. Today, staying true to this spirit, Collins books are packed with inspiration, innovation and practical expertise. They place you at the centre of a world of possibility and give you exactly what you need to explore it.

Collins. Freedom to teach.

Published by Collins
An imprint of HarperCollinsPublishers
77-85 Fulham Palace Road
Hammersmith
London
W6 8JB

Browse the complete Collins catalogue at
www.collinseducation.com

ISBN 978-0-00-726828-3

John Bentham, Colin Chambers, Graham Curtis, Geoffrey Hallas,
Andrew Maczek and David Nicholls assert their moral right to be identified as the authors of this work.

Commissioned by Penny Fowler
Project managed by Alexandra Riley
Edited by Lynn Watkins
Proof read by Patrick Roberts
Design by Newgen Imaging
Cover design by Angela English
Indexing by Laurence Errington
Production by Arjen Jansen
Printed and bound in Hong Kong by Printing Express

540 COL

R7643 2J0502

Contents

3.5.1 Thermodynamics

Enthalpy change, ΔH

The enthalpy change ΔH was introduced in *Collins Student Support Materials: Unit 2 – Chemistry in Action*, section 3.2.1:

> **Definition**
> The **enthalpy change** is the amount of heat taken in or given out at constant pressure during any physical or chemical change.

Essential Notes

A superscript 'plimsoll' sign, $\ominus$, always follows symbols for quantities measured under standard conditions.

Standard enthalpy change, $\Delta H^{\ominus}$

The size of any enthalpy change depends on the *pressure* and the *temperature* as well as on the *amount of substance* used. Chemists use agreed **standard conditions** in order to make useful comparisons between different measurements and between different sets of data.

The **standard enthalpy change**, $\Delta H^{\ominus}$, is defined as follows:

> **Definition**
> The **standard enthalpy change** for a reaction is the change in enthalpy when reactants in their standard states form products that are also in their standard states.

A **standard state** is defined as follows:

> **Definition**
> The **standard state** of a substance at a given temperature is its pure, most stable, form at 100 kPa.

There remains the need to decide both an *amount of substance* and a *temperature* in order fully to specify any enthalpy change, since enthalpy changes can be quoted for any chosen amount of substance, or for any chosen temperature.

Almost always, the **standard amount** used by chemists is the **mole**, as introduced in *Collins Student Support Materials: Unit 1 – Foundation Chemistry*, section 3.1.2. That this standard amount is being used will be perfectly obvious from the units shown, which will almost invariably be quoted as 'per mole' using the symbol **mol^{-1}**.

Examiners' Notes

The most common reference temperature specified is 298 K.

Most frequently, a **reference temperature** of 298 K is used (see *Collins Student Support Materials: Unit 2 – Chemistry in Action*, section 3.2.1). Once this has been specified, ΔH becomes the **standard enthalpy change at 298 K**, shown as $\Delta H^{\ominus}(298\ K)$.

However, it is perfectly in order to quote standard quantities at temperatures different from 298 K, if this is convenient, so $\Delta H^{\ominus}(1000\ K)$ refers to a standard molar enthalpy change at a temperature of 1000 K.

If no temperature is indicated, then it is assumed that the reference temperature is 298 K. Thus, $\Delta H^{\ominus}$ on its own is exactly the same as $\Delta H^{\ominus}$ (298 K).

Some standard enthalpy changes
Some commonly-used standard enthalpy changes are defined below. The standard reference temperature is taken as 298 K in each case.

Standard enthalpy of formation, $\Delta H_f^{\ominus}$
This term was defined in *Collins Student Support Materials: Unit 2 – Chemistry in Action*, section 3.2.1:

> **Definition**
> The **standard enthalpy of formation** is the enthalpy change under standard conditions when one mole of a compound is formed from its elements with all reactants and products in their standard states.

Consequently, the standard enthalpy of formation for an element must always be *zero*. Values of standard enthalpy of formation can be found in data books.

Ionisation enthalpy (energy), $\Delta H_i^{\ominus}$
In *Collins Student Support Materials: Unit 1 – Foundation Chemistry*, section 3.1.1, the *first ionisation energy* of element X was defined as the enthalpy change for the process $X(g) \rightarrow X^+(g) + e^-$. More generally, the **ionisation enthalpy** is defined as follows:

> **Definition**
> The **ionisation enthalpy** is the standard molar enthalpy change for the removal of an electron from a species in the gas phase to form a positive ion and an electron, both also in the gas phase.

This definition embraces all gas phase ionisation processes, not just the **first ionisation enthalpy**, by extending its scope to *any species* (not just to elements), by specifying the *molar quantities* involved, and by referring (correctly, but unimportantly) to electrons in the *gas phase*.

For example, the *ionisation enthalpy* of sodium refers to the process:

$$Na(g) \rightarrow Na^+(g) + e^-(g) \qquad \Delta H_i^{\ominus} = +496 \text{ kJ mol}^{-1}$$

For a neutral species, such as sodium, which loses just a single electron, this is called the first ionisation enthalpy.

The cation formed, $Na^+(g)$, may itself be further ionised, according to the process:

$$Na^+(g) \rightarrow Na^{2+}(g) + e^-(g) \qquad \Delta H_i^{\ominus} = +4562 \text{ kJ mol}^{-1}$$

This is called the **second ionisation enthalpy** of sodium. Some typical values of first and second ionisation enthalpies are shown in Table 1 (see page 6).

Examiners' Notes
The positive ion is called a **cation**; its negative counterpart is an **anion**.

Examiners' Notes
The state symbol (g) for the electron can be assumed and is often omitted.

Examiners' Notes
Enthalpy change is measured at **constant pressure**. Ionisation involves the formation of 2 mol of gaseous products from 1 mol of gaseous reactants, so there is a subtle difference between *ionisation enthalpy* and *ionisation energy*, but this can be ignored here. A similar small difference exists in the case of electron affinity; this too can be ignored.

$\Delta H_i^{\ominus}$ / kJ mol^{-1}		
	First	Second
H(g)	1310	–
He(g)	2370	5250
Mg(g)	736	1450
Na(g)	494	4560

Table 1
Ionisation enthalpy at 298 K

Note that the second ionisation enthalpy of an atom is *always* larger than the first, because the removal of the second electron from an already positively-charged species requires more energy than the removal of the first from a neutral species.

Enthalpy of atomisation, $\Delta H_{at}^{\ominus}$

Definition

The **enthalpy of atomisation** is the standard enthalpy change that accompanies the formation of one mole of gaseous atoms from the element in its standard state.

For an atomic solid, such as an element, **the standard enthalpy of atomisation** is simply the **standard enthalpy of sublimation** of the solid. For example:

$$Na(s) \rightarrow Na(g) \qquad \Delta H_{sub}^{\ominus} = +107 \text{ kJ mol}^{-1}$$

In such a case, the enthalpy of atomisation is identical to the enthalpy of sublimation:

$$\Delta H_{at}^{\ominus} = \Delta H_{sub}^{\ominus}$$

Enthalpies of atomisation (**sublimation**) for a selection of substances are shown in Table 2. Sublimation always requires an input of energy (*endothermic* process), so these enthalpies are all positive.

$\Delta H_{at}^{\ominus}$ / kJ mol^{-1}	
C(graphite)	715
Na(s)	109
K(s)	90
Mg(s)	150

Table 2
Enthalpy of atomisation

In the case of bond fission, a diatomic molecule will produce two moles of atoms, so the enthalpy of atomisation is half the bond dissociation enthalpy (see Table 3). Thus, for chlorine:

$$\tfrac{1}{2} Cl_2(g) \rightarrow Cl(g) \qquad \Delta H_{at}^{\ominus} = \tfrac{1}{2} (+242) = +121 \text{ kJ mol}^{-1}$$

so $\quad \Delta H_{at}^{\ominus} = \tfrac{1}{2} \Delta H_{diss}^{\ominus}$

Bond dissociation enthalpy, $\Delta H_{diss}^{\ominus}$

Definition

The **bond dissociation enthalpy** is the standard molar enthalpy change that accompanies the breaking of a covalent bond in a gaseous molecule to form two gaseous free radicals.

Essential Notes

Homolytic bond fission occurs when a covalent bond breaks and the shared electron pair is split equally between the resulting species (*free radicals*). In **heterolytic** bond fission, both the shared electrons end up on one of the resulting species (*ions*).

In order to indicate that each species formed by bond fission has one unpaired electron, it is considered correct, particularly in processes involving bond fission in a **mass spectrometer** or in **radical chain reactions**, to write a *dot* alongside the odd-electron species to indicate the unpaired electron. This species is called a **free radical**, e.g. the methyl radical •CH_3 or the chlorine atom •Cl.

Definition

A *free radical* is a species that results from the homolytic fission of a covalent bond. It contains an unpaired electron, since homolytic fission results in the splitting of the electron pair in a covalent bond, one electron going to each partner.

In thermodynamic equations involving bond fission, it is quite common to omit *dots* representing unpaired electrons, since their presence is obvious from the equation as written, e.g.

$$Cl_2(g) \rightarrow 2Cl(g) \qquad \Delta H_{diss}^{\ominus} = +242 \text{ kJ mol}^{-1}$$

or

$$CH_4(g) \rightarrow CH_3(g) + H(g) \qquad \Delta H_{diss}^{\ominus} = +435 \text{ kJ mol}^{-1}$$

Some bond dissociation enthalpies for homolytic fission in a selection of covalent compounds are shown in Table 3.

$\Delta H_{diss}^{\ominus}$ / kJ mol^{-1}			
H–H	436	N≡N	945
H–F	565	O=O	496
H–Cl	431	F–F	158
H–Br	366	Cl–Cl	242
H–I	299	Br–Br	194

Table 3
Bond dissociation enthalpy

Electron affinity, $\Delta H_{ea}^{\ominus}$

> **Definition**
>
> **Electron affinity** is the standard enthalpy change when an electron is added to an isolated atom in the gas phase.

Electron affinity refers to a process of the type:

$$Cl(g) + e^-(g) \rightarrow Cl^-(g) \qquad \Delta H_{ea}^{\ominus} = -364 \text{ kJ mol}^{-1}$$

A chlorine atom in the gas phase has a strong **affinity** for an electron, so that the capture of an electron to form a gaseous chloride ion causes energy to be given out to the surroundings – an *exothermic* process.

Some typical values of electron affinity are shown in Table 4.

The enthalpy of formation of $O^{2-}(g)$ from $O^-(g)$ is +844 kJ mol^{-1}. The process is strongly endothermic, since it takes a lot of energy to force a second electron onto the already negative O^- ion.

$\Delta H_{ea}^{\ominus}$ / kJ mol^{-1}	
H(g)	−72
F(g)	−348
Cl(g)	−364
Br(g)	−342
O(g)	−142
O$^-$(g)	+844

Table 4
Electron affinity at 298 K

Lattice enthalpy, $\Delta H_L^{\ominus}$

> **Definition**
>
> **The enthalpy of lattice dissociation** is the standard enthalpy change that accompanies the separation of one mole of a solid ionic lattice into its gaseous ions.
>
> **The enthalpy of lattice formation** is the converse of this, i.e. the standard enthalpy change that accompanies the formation of one mole of a solid ionic lattice from its gaseous ions.

For example:

$$NaCl(s) \rightarrow Na^+(g) + Cl^-(g) \qquad \Delta H_L^{\ominus} = +771 \text{ kJ mol}^{-1}$$

If *lattice dissociation* is used as a defining equation, as above, all lattice enthalpies are *positive*, since all ionic crystals are energetically more favoured than their separated gaseous ions. Consequently, it requires an *input* of energy to disrupt the crystal lattice and form separated gaseous ions.

> **Examiners' Notes**
>
> Bond making and lattice formation are exothermic.
>
> Bond breaking and lattice dissociation are endothermic.

$\Delta H_L^{\ominus}$/ kJ mol^{-1}	
NaF(s)	902
NaCl(s)	771
KCl(s)	701
MgO(s)	3889
MgS(s)	3238

Table 5
Lattice dissociation enthalpy values at 298 K

$\Delta H_{hyd}^{\ominus}$/ kJ mol^{-1}			
Li$^+$	−519	F$^-$	−506
Na$^+$	−406	Cl$^-$	−364
K$^+$	−322	Br$^-$	−335

Table 6
Hydration enthalpy values for individual ions

Examiners' Notes

Enthalpy cycles: A cycle consists of a series of different enthalpy changes which, starting at any point in the cycle, then end up at the same starting point. The sum of all the enthalpy changes round a cycle is zero, $\sum_{cycle} \Delta H = 0$.

If the reverse defining equation is used:

$$Na^+(g) + Cl^-(g) \rightarrow NaCl(s) \qquad \Delta H_L^{\ominus} = -771 \text{ kJ mol}^{-1}$$

then the process of interest is *lattice formation*, and the resulting *enthalpy of lattice formation* is always *negative*. Be aware that such conflicting definitions exist and be able to distinguish which definition is in use from the sign of the resulting lattice enthalpy or the direction of the arrow in the defining equation.

Some typical lattice dissociation enthalpy values are shown in Table 5.

Enthalpy of hydration, $\Delta H_{hyd}^{\ominus}$

Definition

*The **enthalpy of hydration** is the standard enthalpy change for the process:*

$$X^{\pm}(g) \xrightarrow{\text{water}} X^{\pm}(aq) \qquad \Delta H^{\ominus} = \Delta H_{hyd}^{\ominus}$$

where the single symbol, $X^{\pm}(g)$ is used to indicate both a gaseous cation, $X^+(g)$, and a gaseous anion, $X^-(g)$, with $X^{\pm}(aq)$ denoting the appropriate hydrated species in solution.

Hydration enthalpy values for a selection of ions are shown in Table 6.

Enthalpy of solution, $\Delta H_{sol}^{\ominus}$

Definition

*The **enthalpy of solution** is the standard enthalpy change that occurs when one mole of an ionic solid dissolves in enough water to ensure that the dissolved ions are well separated and do not interact with one another.*

For example:

$$NaCl(s) \xrightarrow{\text{water}} Na^+(aq) + Cl^-(aq) \qquad \Delta H_{sol}^{\ominus} = +2 \text{ kJ mol}^{-1}$$

The Born–Haber cycle

The enthalpy of formation of a solid ionic compound can be broken up into a number of steps, arranged in a cycle which is called the **Born–Haber cycle**. Enthalpy cycles were introduced in Unit 2 (*Collins Student Support Materials: Unit 2 – Chemistry in Action*, section 3.2.1); they make use of **Hess's Law** to establish that the sum of the values of all the changes, travelling in either direction round the cycle, must equal zero. This provides a convenient method to find the enthalpy of an unknown step in the cycle if the enthalpy changes for all other steps are known.

The Born–Haber cycle includes a step involving the enthalpy of lattice formation, $\Delta H_L^{\ominus}$, the value of which cannot be determined experimentally, so it is possible to deduce this lattice enthalpy using the Born–Haber cycle and other thermodynamic data. The procedure used is shown in Fig 1.

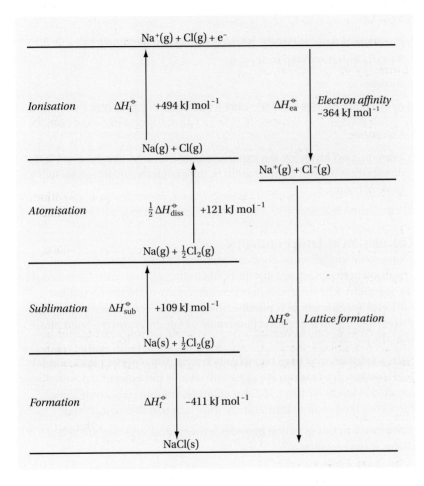

Fig 1
A Born–Haber cycle for the determination of the lattice enthalpy of sodium chloride. The sum of the enthalpy changes round the cycle is zero, so that the magnitude of any unknown enthalpy in the cycle can be found if all the other enthalpies are known

Examiners' Notes

The Born–Haber cycle: starting from the elements, proceed to form the compound by just one step (directly) or by several steps (indirectly).

Examiners' Notes

The term *sublimation* refers to the process in which a solid changes directly into a gas (e.g. $I_2(s) \rightarrow I_2(g)$) without going through an intermediate liquid phase. Note that this term is not included in the AQA A-level specification, but it does provide a useful form of shorthand.

Applying Hess's Law:

$\Delta H^{\ominus}$ (single step) $= \Delta H^{\ominus}$ (five steps)

So $\qquad \Delta H_f^{\ominus} \qquad = +\Delta H_{sub}^{\ominus} + \frac{1}{2}\Delta H_{diss}^{\ominus} + \Delta H_i^{\ominus} + \Delta H_{ea}^{\ominus} + \Delta H_L^{\ominus}$

and $\qquad \Delta H_L^{\ominus} \qquad = -\Delta H_{sub}^{\ominus} - \frac{1}{2}\Delta H_{diss}^{\ominus} - \Delta H_i^{\ominus} - \Delta H_{ea}^{\ominus} + \Delta H_f^{\ominus}$

Thus $\qquad \Delta H_L^{\ominus}(NaCl) = -109 - 121 - 494 - (-364) + (-411)$

$\qquad\qquad\qquad\qquad = -771 \text{ kJ mol}^{-1}$

Examiners' Notes

The value of the standard enthalpy of formation of NaCl(s) used in this calculation is found from data tables and is -411 kJ mol^{-1}.

Example

Calculate the lattice enthalpy of sodium bromide. Use data taken from tables in the text. The enthalpy of vaporisation of bromine is $+30$ kJ mol^{-1} and the standard enthalpy of formation of solid sodium bromide is -360 kJ mol^{-1}.

Method

Construct a cycle, as above. Because bromine is a liquid at 298 K, an extra equation/step is needed.

Answer

$$\Delta H_L^{\ominus}(\text{NaBr}) = -109 - 15 - 97 - 494 + 342 - 360 = -733 \text{ kJ mol}^{-1}$$

Comment

The equation involving the vaporisation of bromine, like the one for the dissociation to form a single atom in the gas phase, involves only half a mole of bromine.

Calculation of lattice enthalpy

Using some fundamental concepts about the forces of attraction and repulsion between ions of like and unlike charges, and with a knowledge of the geometry (the distances between adjacent ions) of the crystalline lattice of an ionic salt, it is possible to use the laws of physics to calculate the energy required to convert one mole of salt ions from the solid phase into the gas phase.

Such a calculation is based on what is known as the **perfect ionic model** and it yields an answer to the question: what is the value of the enthalpy of lattice dissociation of, say, sodium chloride, assuming that only ionic forces hold the sodium ions and chloride ions together in a crystal?

The answer to this question provides a theoretical value for the lattice dissociation enthalpy ΔH_L:

$$\text{NaCl(s)} \rightarrow \text{Na}^+(\text{g}) + \text{Cl}^-(\text{g}) \qquad \Delta H_L(\text{ionic model}) = ?$$

Such a calculated value is based on the assumption that the lattice dissociation enthalpy of an ionic crystal depends solely on the total separation of ions that experience only purely ionic (electrostatic) forces. This ionic model makes it possible to allow a comparison between:

- an experimentally-based value of the lattice dissociation enthalpy (found by using a Born–Haber cycle) and

- a theoretically-based value, calculated by assuming that ionic (electrostatic) forces are solely responsible for the stability of ionic crystalline solids.

Examiners' Notes

The questions raised here, as well as their answers, provide a graphic illustration of *How Science Works*. Data from one source (experiment + Born–Haber) are compared with potentially equally valid data from another source (perfect ionic model theory + calculation). Agreement confers a degree of validity on the assumptions made in the theoretical model; disagreement provides a motive and a pressing need to ask further questions and to seek alternative explanations.

With these two methods (experiment and calculation) available, the first thing to do is to see if the results agree and, if so, how well and for which compounds. Given the approximations that are needed in making valid calculations, agreement needs only to be close, not exact.

The most relevant first point of comparison between experiment and calculation is for the alkali metal halides. Table 7 shows values of experimental *Born–Haber* (*BH*) lattice dissociation enthalpies alongside the corresponding calculated *ionic model* (*IM*) values.

ΔH_L / kJ mol^{-1}								
	F		Cl		Br		I	
	BH	IM	BH	IM	BH	IM	BH	IM
Li	1022	1004	846	833	800	787	744	728
Na	902	891	771	766	753	752	684	686
K	801	795	701	690	670	665	629	632

Table 7
Experimental (*BH*) and calculated (*IM*) lattice dissociation enthalpies for some alkali metal halides

There is very satisfactory agreement between the Born–Haber (experimental) values and the perfect ionic model (theoretical) values in this Table. The average discrepancy amounts to just over 8 kJ mol^{-1}, which is roughly of the order of 1%. It is therefore reasonable to conclude that these simple alkali halide ionic crystals behave just as would be expected for systems consisting of oppositely-charged discrete ions interacting exclusively through electrostatic attractions and repulsions. The ionic model works almost perfectly.

However, such good agreement does not persist for other presumed ionic crystals. Table 8 contains data for the silver halides.

ΔH_L / kJ mol^{-1}								
	F		Cl		Br		I	
	BH	IM	BH	IM	BH	IM	BH	IM
Ag	955	870	905	770	890	758	876	736

Table 8
Experimental (*BH*) and calculated (*IM*) lattice dissociation enthalpies for the silver halides

Here, there is far less agreement. The average discrepancy amounts to just over 120 kJ mol^{-1}, or some 13–14%. Some other factor must be having an influence.

A simple explanation of these discrepancies is that the ionic model takes no account of such factors as a degree of **covalent bonding** between the constituent ions. Such additional bonding would require an additional input of energy in order to separate the constituents of the solid into free

Examiners' Notes

Experimental values of the lattice dissociation enthalpy seem almost always to be greater than those calculated using the *perfect ionic model*. This discrepancy can be accounted for if it is assumed that some presumed ionic compounds possess a degree of covalent bonding (*covalent character*) which adds stability to the solid state and thus requires additional energy in converting the solid into independent gaseous ions.

separate gaseous ions. This requirement is consistent with the considerably greater value of the experimental lattice dissociation enthalpy (*BH*) as compared with the purely ionic calculation (*IM*).

The increased lattice enthalpy of the silver halides compared with the ionic model may account for their lack of solubility in water. It might be expected that silver chloride, were it a simple ionic compound like sodium chloride, would be soluble in water. However, the additional heat energy required to separate the ions before they are hydrated makes the overall free-energy change for this process positive (see page 19), so that dissolving is no longer a 'feasible' process.

Calculating enthalpies of solution

The following example shows a calculation that uses enthalpies of hydration and lattice enthalpy.

Example

Calculate the enthalpy of solution of sodium chloride, using data taken from tables in the text.

Answer

Method 1

Draw an enthalpy cycle as shown below.

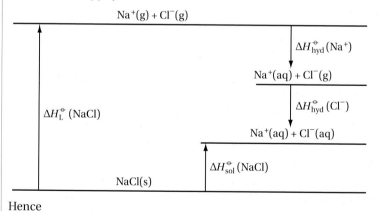

Hence

$$\Delta H_{sol}^{\ominus} \text{ (NaCl)} = \Delta H_{L}^{\ominus} \text{ (NaCl)} + \Delta H_{hyd}^{\ominus} \text{ (Na}^+\text{)} + \Delta H_{hyd}^{\ominus} \text{ (Cl}^-\text{)}$$

Therefore

$$\Delta H_{sol}^{\ominus} \text{ (NaCl)} = \quad +771 \quad + \quad (-405) \quad + \quad (-364) \quad = +2 \text{ kJ mol}^{-1}$$

Examiners' Notes

The arrow in the diagram leads from solid NaCl to separated gaseous ions. Clearly, $\Delta H_{L}^{\ominus}$ (NaCl) in this case is the *enthalpy of lattice dissociation*.

Method 2

Write equations, including enthalpy changes (reversed where appropriate) for lattice enthalpy and hydration enthalpies in such a way that simple addition yields the equation for enthalpy of solution. Addition of the enthalpy changes gives the overall enthalpy of reaction.

$$NaCl(s) \rightarrow Na^+(g) + Cl^-(g) \qquad \Delta H_L^\ominus \quad = +771 \text{ kJ mol}^{-1}$$

$$Na^+(g) \xrightarrow{\text{water}} Na^+(aq) \qquad \Delta H_{hyd}^\ominus = -405 \text{ kJ mol}^{-1}$$

$$Cl^-(g) \xrightarrow{\text{water}} Cl^-(aq) \qquad \Delta H_{hyd}^\ominus = -364 \text{ kJ mol}^{-1}$$

$$NaCl(s) \xrightarrow{\text{water}} Na^+(aq) + Cl^-(aq) \quad \Delta H_{sol}^\ominus \quad = +2 \text{ kJ mol}^{-1}$$

Comment

Either method may be used with confidence. The one that is chosen is entirely a matter of personal preference, although *Method 2* is slightly more transparent and marginally less prone to mistakes.

Note also that the enthalpy of solution is *very* small, and is calculated here as the difference between two very large quantities. Such calculations are often very unreliable and are subject to round-off errors, particularly where other enthalpies are quoted to only three significant figures. Their difference, therefore, is good only to one significant figure.

Mean bond enthalpies

The concept of **mean bond enthalpy** was introduced in *Collins Student Support Materials: Unit 2 – Chemistry in Action,* section 3.2.1, pages 8–9. This section recaps and extends the information found in that section.

When a reaction occurs, bonds in molecules are broken (requiring an input of heat energy from the surroundings) and new bonds are formed (causing a flow of heat energy into the surroundings). The overall enthalpy change for the reaction represents the net balance between heat energy taken in and heat energy evolved; it is the balance between old bonds broken and new bonds formed.

The strengths of bonds can be determined using spectroscopy, by measuring the **bond dissociation enthalpy**, $\Delta H_{diss}^\ominus(X\text{—}Y)$, introduced earlier. This is the enthalpy change accompanying the fission of a bond in a gas-phase species, X—Y(g), to form two new gas-phase fragments:

$$X\text{—}Y(g) \rightarrow X(g) + Y(g) \qquad \Delta H^\ominus = \Delta H_{diss}^\ominus(X\text{—}Y)$$

If all **bond enthalpies** were known, it would be possible to predict overall enthalpy changes simply by summing over those bonds that had altered – writing a positive enthalpy contribution for any bond broken, and a negative one for any bond formed.

Unfortunately such a simplistic approach fails. It is clear why this is so if the case of water is considered. Water has two entirely equivalent O—H bonds. Yet, breaking the first:

$$H_2O(g) \rightarrow H(g) + OH(g) \qquad \Delta H_{diss}^\ominus(H\text{—}OH) = +492 \text{ kJ mol}^{-1}$$

Examiners' Notes

The *dissociation enthalpy* of a given bond (O—H, say) depends on the structure of the rest of the molecule around it. To break an O—H bond in ethanol requires a different input of heat energy from that needed to break such a bond in ethanoic acid.

Examiners' Notes

The *mean bond enthalpy* of a bond X—Y is given the symbol $\Delta H_B(X—Y)$.
It is an *average* value and is useful in *approximate* calculations.

gives rise to a bigger enthalpy change than breaking the second:

$$HO(g) \rightarrow H(g) + O(g) \qquad \Delta H_{diss}^{\ominus} (O—H) = +428 \text{ kJ mol}^{-1}$$

Changes in the environment of the bond change its strength. In methanol, for example:

$$CH_3OH(g) \rightarrow CH_3O(g) + H(g) \qquad \Delta H_{diss}^{\ominus} (CH_3O—H) = +437 \text{ kJ mol}^{-1}$$

the bond enthalpy corresponds to the breaking of neither of the two O—H bonds in water.

The solution to this difficulty lies in compiling a list of mean bond enthalpies, $\Delta H_B(X—Y)$. The mean bond enthalpy is defined as follows:

Definition

The **mean bond enthalpy** is the average of several values of the bond dissociation enthalpy for a given type of bond, taken from a range of different compounds.

$\Delta H_B(X—Y) / \text{kJ mol}^{-1}$			
H–H	436	C–Cl	328
H–C	412	C–N	305
H–N	391	C–O	360
H–O	463	C=O	743
C–C	348	N–O	158
C=C	612	O–O	146
C≡C	838	O=O	496

Table 9
Mean bond enthalpies

By averaging bond dissociation enthalpies measured in a wide variety of different compounds, mean values can be found. Some values are shown in Table 9.

Mean bond enthalpies provide a simple way of calculating approximate values of overall enthalpy changes, particularly where other data are not available. However, the values obtained are only *approximate* and should always be used with caution. An illustration of the use of mean bond enthalpies is given in the following example.

Example

Calculate the standard enthalpy change for the complete combustion of propane. Use the mean bond enthalpy data given in Table 9.

Method

Firstly, write an equation for the overall reaction. Next, '*atomise*' the reactant molecules by breaking all the bonds present on the reactant side of the equation.

Then '*reassemble*' the product molecules from the gaseous atoms by making all the bonds present on the product side of the equation. The overall enthalpy change is the sum of the enthalpy needed to break the bonds and the enthalpy released when bonds are made.

Enthalpy change = energy required for breaking bonds

+ energy released from making bonds

Answer

The overall equation is:

$$C_3H_8(g) + 5O_2(g) \rightarrow 3CO_2(g) + 4H_2O(g)$$

Examiners' Notes

Bond breaking is an *endothermic* process (ΔH is positive).

Bond making is an *exothermic* process (ΔH is negative).

Draw an enthalpy cycle as shown below.

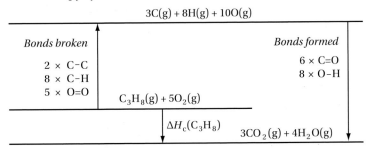

$$3C(g) + 8H(g) + 10O(g)$$

Bonds broken	Bonds formed
$2 \times$ C–C	$6 \times$ C=O
$8 \times$ C–H	$8 \times$ O–H
$5 \times$ O=O	

$$C_3H_8(g) + 5O_2(g)$$

$\Delta H_c(C_3H_8)$

$$3CO_2(g) + 4H_2O(g)$$

Bonds broken/kJ mol^{-1}	Bonds formed/kJ mol^{-1}
$2 \times \Delta H_B$(C–C) $= 2 \times 348 = 696$	
$8 \times \Delta H_B$(C–H) $= 8 \times 413 = 3304$	$6 \times \Delta H_B$(C=O) $= 6 \times 743 = 4458$
$5 \times \Delta H_B$(O=O) $= 5 \times 496 = 2480$	$8 \times \Delta H_B$(O–H) $= 8 \times 463 = 3704$

Energy taken in = 6480   Energy given out = 8162

$$\Delta H_c(C_3H_8) = \Sigma \Delta H_B \text{(bonds broken)} - \Sigma \Delta H_B \text{(bonds formed)}$$

$$= \quad 6480 \quad\quad - \quad\quad 8162$$

So $\Delta H_c(C_3H_8) = -1682$ kJ mol^{-1}

Comment

The value calculated using mean bond enthalpies is only *approximate*, because the *mean bond enthalpy data* used refer to *average* values of bond enthalpies derived from measurements in many different compounds.

Examiners' Notes

The value found using mean bond enthalpies is often within 10% of the correct (*calorimetric*) value, which is typical of the level of accuracy that can be expected from this method. In cases where calorimetric data are not available, *mean bond enthalpy* values can provide a satisfactory estimate of inaccessible reaction enthalpy data.

In many reactions, some of the bonds broken in complete *atomisation* are immediately reformed in *reassembly*, thus rendering much of the breaking and reforming a self-cancelling exercise. In such cases, it is perfectly legitimate to count only the bonds broken that are not reformed. For example, in the case of the oxidation by gaseous oxygen of propan-1-ol to propanoic acid, the overall equation is:

$$CH_3CH_2CH_2OH + O_2 \rightarrow CH_3CH_2COOH + H_2O$$

and the only bonds broken and formed are:

$$2 \times \text{C–H} + 1 \times \text{O=O} \rightarrow 1 \times \text{C=O} + 2 \times \text{O–H}$$

Adding and subtracting six mean bond enthalpies (as above) is a much more sensible process than including all of the extra 9 broken and then re-formed bonds as would be demanded by complete atomisation.

Examiners' Notes

It saves time and makes sense to include only the new bonds formed from the old bonds broken in any mean bond enthalpy determination of ΔH.

Free-energy change (∆G) and entropy change (∆S)

Spontaneous change

> ### Definition
> A **spontaneous change** is one that has a natural tendency to occur without being driven by any external influences.

Spontaneous changes are familiar from everyday life. The air compressed in a bicycle tyre escapes spontaneously if the valve is removed, and a lot of physical effort is needed to pump it up again. Hot soup cools spontaneously, and heat energy from a gas ring is needed to warm it up again. Ice cream melts spontaneously on a hot day, and needs the electrical energy put into a refrigerator to freeze it again. Iron rusts spontaneously in damp air, and it takes all the vast energy of a blast furnace to get iron back from an oxide ore. A rechargeable dry cell can light the bulb of a torch spontaneously, but has to be left to draw energy from the mains when being recharged.

Left to themselves, therefore, some things simply happen. Others do not happen unless energy is expended. A **spontaneous change** is one that can occur in one particular direction but not in reverse (unless conditions such as temperature are changed).

The enthalpy factor

Many spontaneous chemical reactions appear to be driven by a favourable change in enthalpy. In all the changes listed above, it would be easy to reach the simple (but mistaken) conclusion that spontaneous change occurs if heat is given out as the change is taking place. But this is not the whole story.

Exothermic reactions are often spontaneous. The natural direction of spontaneous change is from higher to lower enthalpy, with a release (exothermic) of the difference in energy between the two. For example, the reaction:

$$H_2(g) + \tfrac{1}{2}O_2(g) \rightarrow H_2O(g) \qquad \Delta H^{\ominus} = -242 \text{ kJ mol}^{-1}$$

liberates vast quantities of heat energy. The reaction is exothermic, the enthalpy change is negative ($\Delta H < 0$), and this favours spontaneous reaction.

It would be tempting to conclude that spontaneous reactions must involve a release of heat energy. However, some reactions are spontaneous even though they are **endothermic**. For example, the reaction:

$$KHCO_3(s) + HCl(aq) \rightarrow KCl(aq) + H_2O(l) + CO_2(g) \qquad \Delta H^{\ominus} = +25 \text{ kJ mol}^{-1}$$

is endothermic and the temperature of the reaction mixture drops when the reactants are mixed. Yet it proceeds spontaneously. Clearly, there must be some additional factor that causes reactions to occur, over and above the simple release of heat energy.

The entropy factor

The additional factor that helps to drive spontaneous change in a given direction is called the **entropy**, which is given the symbol *S*.

Before proceeding, in the next sections, to relate entropy to everyday experience, it is useful to make a few statements about entropy and its

Examiners' Notes

If a process in one direction is spontaneous, the reverse process will require an input of external energy.

Examiners' Notes

Most spontaneous reactions are exothermic, but only some endothermic reactions are spontaneous.

Essential Notes

Entropy (*S*) has the units $J \text{ K}^{-1} \text{ mol}^{-1}$.
Note that J rather than kJ is used.

nature, just to get an insight into this new quantity that turns out to be one of the driving forces in spontaneous change.

(a) Entropy and disorder

It proves very helpful to *think of* entropy in terms of *disorder*. An increase in entropy can be visualised as an increase in disorder. Processes leading to increased chaos are rather more likely than those leading to order. Heating leads to an increase in disorder, which fits with the increase in entropy as a substance changes from solid, to liquid, to gas.

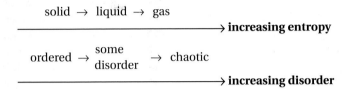

(b) Variation of entropy with temperature

A graph showing typical changes of entropy as temperature is increased is shown below.

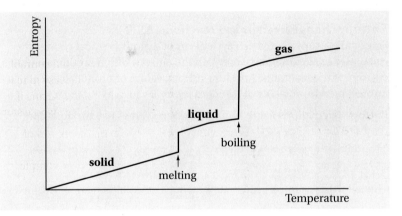

It is clear that:

- entropy increases slowly with temperature in solids, liquids and gases
- a **phase change** (*change of physical state*) causes a sudden change in entropy
- boiling causes a larger increase in entropy than does melting.

(c) Magnitude of entropy in different substances

In Table 10, some **standard entropy** values of different substances are listed. Two broad trends can be noted:

- simple molecules generally seem to have lower entropies than more complicated molecules, for example:

$$S^{\ominus}(CO) < S^{\ominus}(CO_2); \qquad S^{\ominus}(He) < S^{\ominus}(O_2); \qquad S^{\ominus}(NaCl) < S^{\ominus}(NaHCO_3)$$

- for substances of similar complexity, entropy generally increases on going from solid, to liquid, to gas, for example:

$$S^{\ominus}(H_2O(s)) < S^{\ominus}(H_2O(l)) < S^{\ominus}(H_2O(g)); \qquad S^{\ominus}(CH_3OH(l)) < S^{\ominus}(CH_3OH(g))$$

$S^{\ominus}_{298}$ / J K^{-1} mol^{-1}			
diamond	2.4	He(g)	126
graphite	5.7	Ar(g)	155
NaCl(s)	72	O_2(g)	205
NaHCO$_3$(s)	102	CO(g)	198
SiO$_2$(s)	42	CO$_2$(g)	214
H$_2$O(l)	70	H$_2$O(g)	189
CH$_3$OH(l)	127	CH$_3$OH(g)	240

Table 10
Standard entropy at 298 K

(d) Units in which S is measured

In the SI system, the units of entropy are J K^{-1} mol^{-1}. In themselves, these units say little that is new about entropy, but they will prove useful and revealing later on when the separate influences of enthalpy and entropy are combined in determining the direction of spontaneous change.

(e) Absolute entropy, $S^{\ominus}$

Since entropy is so closely linked to disorder, and as it decreases when the temperature is lowered, it is reasonable to suggest that, at 0 K (the absolute zero), all disorder will vanish and all substances will be perfectly ordered and have *zero entropy*. This conclusion turns out to be true for most substances, and especially for perfectly-ordered crystals.

Examiners' Notes

Absolute entropy values are based on the premise that all substances have zero entropy at the absolute zero of temperature.

The concept of zero entropy has one important consequence. With enthalpy, it is possible to consider only standard enthalpy *changes*, $\Delta H^{\ominus}$ (*i.e. only differences between enthalpy values*); in the case of entropy, there is a well-defined starting point to the entropy scale (zero entropy at zero Kelvin). As a consequence, it is possible to speak of and to list values for *absolute standard entropies, $S^{\ominus}$*, and from these to calculate *standard entropy changes, $\Delta S^{\ominus}$*. Table 10 lists some **absolute entropy values.**

Entropy change in chemical reactions, $\Delta S^{\ominus}$

Standard entropy changes in systems undergoing chemical change are calculated as the difference between the entropy of the products and the entropy of the reactants. Absolute entropy values of products and reactants are obtained from tables of *standard entropies* such as Table 10.

Entropy calculations using standard entropy values use the equation:

$$\Delta S^{\ominus} = \Sigma S^{\ominus}_{\text{products}} - \Sigma S^{\ominus}_{\text{reactants}}$$

The following examples illustrate how such calculations are performed.

Example

Calculate the entropy change that accompanies the combustion of graphite. Use the entropy data in Table 10.

Method

The reaction in question is:

$$C(s) + O_2(g) \rightarrow CO_2(g)$$

Answer

$$\Delta S^{\ominus} = \Sigma S^{\ominus}_{\text{products}} - \Sigma S^{\ominus}_{\text{reactants}}$$

$\Sigma S^{\ominus}_{\text{products}}$	= 214.0 J K^{-1} mol^{-1}
$\Sigma S^{\ominus}_{\text{reactants}}$ = 5.7 + 205	= 210.7 J K^{-1} mol^{-1}
So $\Delta S^{\ominus}$ = 214.0 − 210.7	= +3.3 J K^{-1} mol^{-1}

Comment

The entropy change here is quite small. Carbon dioxide and oxygen are both gases and have similar entropies. Graphite is a very low-entropy solid. So even though two moles form one, the number of gas moles does not change and there is not much change in entropy.

Example

Calculate the entropy change that accompanies the decomposition by heating of solid sodium hydrogencarbonate. Use the entropy data given in Table 10, and the additional information that solid sodium carbonate has a standard entropy of 136 J K^{-1} mol^{-1}.

Method

The reaction in question is:

$$2NaHCO_3(s) \rightarrow Na_2CO_3(s) + H_2O(g) + CO_2(g)$$

Answer

$$\Delta S^\ominus = \Sigma S^\ominus_{products} - \Sigma S^\ominus_{reactants}$$

$$\Sigma S^\ominus_{products} = 136 + 189 + 214 = 539 \text{ J K}^{-1} \text{ mol}^{-1}$$

$$\Sigma S^\ominus_{reactants} = 2 \times 102 \qquad = 204 \text{ J K}^{-1} \text{ mol}^{-1}$$

So $\Delta S^\ominus \quad = 539 - 204 \qquad = +335 \text{ J K}^{-1} \text{ mol}^{-1}$

Comment 1

The entropy change calculated is *per mole of equation as written,* which here refers to the decomposition of 2 mol of NaHCO$_3$. For one mole of NaHCO$_3$, ΔS is halved, giving 167.5 J K^{-1} mol^{-1}.

Comment 2

The entropy change here is very much larger than that found in the previous example. In the case above, 2 mol of gas are formed from a solid, so the disorder of the system is increased much more.

Gibbs free-energy change, $\Delta G^\ominus$, spontaneous change and feasibility

It is clear that reactions in which there is a release of energy (**exothermic processes**) tend to happen quite often. Equally, reactions in which there is an intake of energy (**endothermic processes)** but which lead to an increase in disorder (*increase in entropy*) also happen quite often. The sometimes conflicting demands of enthalpy and entropy are brought together in the relationship:

$$\Delta G^\ominus = \Delta H^\ominus - T\Delta S^\ominus$$

where the standard free-energy changes, $\Delta G^\ominus$, combines the influence of both enthalpy and entropy changes.

Essential Notes

The unit mol^{-1} (*'per mole'*) means *'per mole of the defining equation, as written'.* If the defining equation is doubled, then all the associated thermodynamic quantities are also doubled.
All *molar* thermodynamic quantities are tied, uniquely, to their own defining equations.

This equation expresses the benefit of an exothermic reaction (ΔH *negative*) with one in which entropy increases (ΔS *positive*); both of these conditions are associated with changes that do happen, leading to the conclusion that:

ΔG must be negative (or zero) for spontaneous change.

For example, the reaction between gaseous hydrogen and oxygen at room temperature is classed as **spontaneous** (or **feasible**) because ΔG for the formation of water from its elements is very negative. So it *can* go. The reaction *will not* go, however, unless a spark or a flame is applied, whereupon an explosion results.

Similarly, hydrogen and chlorine do not react in the dark, but do so with explosive violence in sunlight.

In such cases, the reaction needs to overcome an **activation energy** barrier (see *Collins Student Support Materials: Unit 2 – Chemistry in Action*, section 3.2.2) before reaction can occur.

Spontaneous ≡ feasible

In thermodynamics, the words *spontaneous* and *feasible* have exactly the same meaning. In everyday speech, however, the word *spontaneous* (as in *spontaneous applause* or *spontaneous tears*) implies something that is almost *inevitable*, that simply *must* happen. In thermodynamics, spontaneity has to do only with a *tendency* for something to happen, and not at all if it *actually will* happen. The term '*feasible*' is much to be preferred, implying as it does, something that is possible but not inevitable, and it is the term that will be used in what now follows. However, it is worth noting that the term '*spontaneous*' is firmly enshrined in the language of equilibrium and that it frequently provides an acceptable alternative to *feasible*.

Quite often, the speed at which a feasible change occurs can be rather slow (as in the cooling of a cup of coffee), or even infinitely slow (as in the change from diamond to graphite). Nonetheless, both these changes (cold coffee and no diamonds) are truly *feasible* because a decrease in free energy occurs.

Thermodynamics has nothing to say about the speed with which things happen: only whether they *can* happen. Thermodynamics answers the first and fundamental question: *can it go?* (is the reaction *feasible*, is $\Delta G < 0$?). Only if the answer to this first question is '*yes*' is it worth posing the second question: *how fast will it go?* (are the *kinetics* favourable?).

A chemical or physical change is said to be *feasible* if the value of ΔG is *negative* or *zero*. When ΔG is *positive*, the change is said to be *unfeasible* (or *not feasible*).

The way to think of *feasibility*, and specifically of the criterion $\Delta G = 0$, is to understand that $\Delta G = 0$ defines a break-even point where, on balance, there will be similar concentrations of reactants and products present at equilibrium. If $\Delta G < 0$, there will be more products than reactants at equilibrium; if $\Delta G > 0$, reactants will predominate.

Examiners' Notes

A relatively small change in the value of ΔG can tip equilibrium from being largely favourable to being rather unfavourable. At 300 K, as little as a ±6 kJ mol^{-1} change in ΔG can tip an equilibrium from 10:1 in favour of reactants to 10:1 in favour of products.

Calculations involving $\Delta G^{\ominus}$

From standard enthalpy and standard entropy data, the standard free-energy change can be calculated using $\Delta G = \Delta H - T\Delta S$, from which the feasibility of a chemical reaction can be determined. This procedure is illustrated in the examples that follow. The data required are given in Table 11.

Chemical reaction	$\Delta H^{\ominus}_{298}$ / kJ mol^{-1}	$\Delta S^{\ominus}_{298}$ / J K^{-1} mol^{-1}
$C(s) + O_2(g) \rightarrow CO_2(g)$	−394	+3.3
$2Fe(s) + \frac{3}{2}O_2(g) \rightarrow Fe_2O_3(s)$	−825	−272
$2NaHCO_3(s) \rightarrow Na_2CO_3(s) + H_2O(g) + CO_2(g)$	+130	+335

Table 11
Standard enthalpy and entropy changes for three different chemical reactions

Examiners' Notes

Note that the first reaction is always feasible, that low temperatures favour the second, while the third is favoured by high temperatures.

Example

Calculate the standard free-energy change for the combustion of graphite at 298 K. Use the data given in Table 11.

The reaction in question is:

$$C(s) + O_2(g) \rightarrow CO_2(g)$$

Answer

$$\Delta G^{\ominus} = \Delta H^{\ominus} - T\Delta S^{\ominus}$$

$$\Delta G^{\ominus}_{298} = -394 \text{ kJ mol}^{-1} - \frac{298 \times 3.3}{1000} \text{ kJ mol}^{-1}$$

$$= -395 \text{ kJ mol}^{-1}$$

Comment

The energy units of ΔH and $T\Delta S$ must be made compatible using division by 1000 to convert the value of ΔS (given in J K^{-1} mol^{-1}) into kJ K^{-1} mol^{-1}. This operation renders the entropy term (3.3×10^{-3} kJ K^{-1} mol^{-1}) vanishingly small, even when multiplied by 298 K.

Examiners' Notes

ΔG is negative, so the reaction is *feasible*. The entropy term is very small so $\Delta G \cong \Delta H$. Because ΔS is so small, ΔG will not vary much with temperature, though in practice the reaction is extremely slow at 298 K (high activation energy).

Example

Calculate the standard free-energy change for the rusting of iron at 298 K. Use the data given in Table 11.

The reaction in question is:

$$2Fe(s) + \frac{3}{2}O_2(g) \rightarrow Fe_2O_3(s)$$

Answer

$$\Delta G^{\ominus} = \Delta H^{\ominus} - T\Delta S^{\ominus}$$

$$\Delta G^{\ominus}_{298} = -825 \text{ kJ mol}^{-1} - \frac{298 \times (-272)}{1000} \text{ kJ mol}^{-1}$$

$$= -744 \text{ kJ mol}^{-1}$$

Comment

ΔG is hugely negative, so that the rusting of iron is highly feasible at room temperature, in complete accord with everyday experience. The reaction is favoured by low temperatures but does not cease to be feasible until the temperature exceeds 3000 K. It is strongly driven by a very exothermic enthalpy change.

Examiners' Notes

ΔG here is very negative, so the reaction is *feasible*. The entropy change is negative because disorder in the gas is lost forming a solid. ΔG is negative, however, because the exothermic value of ΔH dominates.

Example

Calculate the standard free-energy change for the decomposition of one mole of sodium hydrogencarbonate at 298 K. Use the data in Table 11.

The reaction in question is:

$$2NaHCO_3(s) \rightarrow Na_2CO_3(s) + H_2O(g) + CO_2(g)$$

Answer

$$\Delta G^\ominus = \Delta H^\ominus - T\Delta S^\ominus$$

$$\Delta G^\ominus = 130 \text{ kJ mol}^{-1} - \frac{298 \times 335}{1000} \text{ kJ mol}^{-1} = +30 \text{ kJ mol}^{-1}$$

So, for 1 mol of $NaHCO_3$, $\Delta G^\ominus_{298} = \frac{1}{2} \times 30 = +15 \text{ kJ mol}^{-1}$ (*2 s.f.*)

Comment 1

The unit mol^{-1} means *per mole of the equation as written* which, in this case, involves 2 mol of $NaHCO_3$.

Comment 2

ΔG is only slightly positive at 298 K, so the decomposition of sodium hydrogencarbonate is *almost* feasible at room temperature, but needs some heat input to become *actually* feasible. Again, this is in accord with everyday experience. The reaction is favoured by high temperatures and needs a temperature close to 400 K to become feasible (see the example on pages 24 and 25).

Examiners' Notes

ΔG is positive, so at 298 K the reaction is *not feasible*. However, ΔG is not very large, so raising the temperature will allow the $-T\Delta S$ term to dominate. The reaction will become *feasible* at higher temperatures. The entropy change is positive since one gaseous mole forms for the loss of half a mole of solid.

Entropy in physical changes

In many physical changes, there is an increase or decrease in order (and hence in entropy). The most obvious of such changes is the melting of a solid or the boiling of a liquid, and these two cases are considered in the examples shown on pages 23 and 24, respectively.

Melting (fusion)

When ice melts at constant temperature (0 °C) and constant pressure (100 kPa), the mixture of ice and water that results has *no spontaneous tendency* either to solidify (become all ice) or to liquefy (become all water) unless external influences (addition or removal of heat) are involved. If left in a system which prevents a flow of heat (e.g. a thermos flask) an ice/water mixture will remain at 0 °C as long as both ice and water are present. If a little heat leaks in, a little of the ice absorbs this heat and melts. The new equilibrium has slightly different amounts of ice and water present, but the temperature remains at 0 °C.

This is an example of a **system at equilibrium** that has no tendency to move spontaneously in one direction or the other.

Since it is the sign of ΔG that determines the tendency of a system to change, it can be concluded that:

For a system at equilibrium, $\Delta G = 0$

Applying this criterion, $\Delta G = 0$, to the equation $\Delta G^{\ominus} = \Delta H^{\ominus} - T\Delta S^{\ominus}$ leads to:

$$\Delta H^{\ominus} = T\Delta S^{\ominus}$$

So, for melting *(fusion)*, $\Delta S_{fus}^{\ominus} = \dfrac{\Delta H_{fus}^{\ominus}}{T_{fus}}$

Boiling (vaporisation)

Arguments similar to the ones used for melting also apply here.

When water boils at constant temperature (100 °C) and constant pressure (100 kPa), the mixture of water and steam that results has *no spontaneous tendency* either to liquefy (become all water) or to vaporise (become all steam) unless external influences (addition or removal of heat) are involved. Left in a system which prevents a flow of heat, a water/steam mixture will remain at 100 °C as long as both water and steam are present. If a little heat leaks out, a little of the steam condenses and replaces the lost heat. The new equilibrium will have slightly different amounts of water and steam present, but the temperature will remain at 100 °C.

This is also a system at equilibrium that has no tendency to move spontaneously in one direction or the other.

Again, it is the sign of ΔG that determines the tendency of a system to change, so it can be concluded that:

for boiling *(vaporisation)*, $\Delta S_{vap}^{\ominus} = \dfrac{\Delta H_{vap}^{\ominus}}{T_{vap}}$

The following examples illustrate the entropy changes that accompany the melting of ice and the boiling of water.

Example

Calculate the entropy change that accompanies the melting of ice. The enthalpy of fusion of ice is 6.0 kJ mol^{-1}.

Method

The reaction in question is:

$$H_2O(s) \rightleftharpoons H_2O(l)$$

This is an equilibrium change of state at fixed temperature, so $\Delta G = 0$.

$$\Delta S_{fus}^{\ominus} = \dfrac{\Delta H_{fus}^{\ominus}}{T_{fus}}$$

Answer

$$\Delta H_{fus}^{\ominus} = +6.0 \text{ kJ mol}^{-1}$$
$$T_{fus} = 273 \text{ K}$$

Therefore $\quad \Delta S_{fus}^{\ominus} = \dfrac{+6.0 \times 10^3 \text{ J mol}^{-1}}{273 \text{ K}} = +22 \text{ J K}^{-1} \text{ mol}^{-1}$

Comment

ΔS is positive, as is to be expected when a solid melts to form a liquid. The only difficulty worth noting is the need to convert enthalpy (in kJ mol^{-1}) into J mol^{-1} in order to give entropy in J K^{-1} mol^{-1}.

Examiners' Notes

The influence on the *position of equilibrium* of the addition or removal of heat to melting ice provides a graphic example of *Le Chatelier's principle* in action. The system responds to nullify the influence of the heat added or removed, and does so very successfully, maintaining the temperature at a constant 0 °C.

Examiners' Notes

Another example of *Le Chatelier's principle* in action.
When both phases are present in equilibrium at constant pressure, the temperature remains constant.
This property of equilibria is used in establishing two fixed points (0 °C and 100 °C) on the Celsius temperature scale.

Example

Calculate the entropy change that accompanies the boiling of water. The enthalpy of vaporisation of water is +44.0 kJ mol^{-1}.

Method

The reaction in question is:

$$H_2O(l) \rightleftharpoons H_2O(g)$$

This is an equilibrium change of state at fixed temperature, so $\Delta G = 0$.

$$\Delta S^{\ominus}_{vap} = \frac{\Delta H^{\ominus}_{vap}}{T_{vap}}$$

Answer

$$\Delta H^{\ominus}_{vap} = +44.0 \text{ kJ mol}^{-1}$$
$$T_{vap} = 373 \text{ K}$$

Therefore
$$\Delta S^{\ominus}_{vap} = \frac{+44.0 \times 10^3 \text{ J mol}^{-1}}{373 \text{ K}} = +118 \text{ J K}^{-1} \text{ mol}^{-1}$$

Comment

The entropy change here has a much more positive value than that for a solid melting to form a liquid. The increase in entropy is associated with the creation of one mole of disordered vapour molecules from one mole of relatively ordered liquid molecules.

Calculation of the temperature at which a reaction becomes feasible

A reaction that is not feasible at one temperature may become feasible if the temperature is raised (or lowered). The temperature at which feasibility is just achieved is the temperature at which there is no spontaneous tendency for the reaction to go either one way or the other. At this temperature, ΔG becomes zero and

$$\Delta H^{\ominus} = T\Delta S^{\ominus}$$

So, for feasibility,
$$T = \frac{\Delta H^{\ominus}}{\Delta S^{\ominus}}$$

The **temperature of feasibility** can then be calculated if ΔH and ΔS are both known. An illustration of such a calculation is given in the example shown below.

Example

Calculate the temperature at which the thermal decomposition of sodium hydrogencarbonate becomes feasible. The standard enthalpy and entropy changes for the decomposition are given in Table 11.

The reaction in question is:

$$2NaHCO_3(s) \rightleftharpoons Na_2CO_3(s) + H_2O(g) + CO_2(g)$$

Answer

The forward reaction becomes feasible when $\Delta G = 0$.

Therefore,

$$\Delta H^{\ominus} = T\Delta S^{\ominus} \quad \text{and} \quad T = \frac{\Delta H^{\ominus}}{\Delta S^{\ominus}}$$

So $\quad T = \dfrac{+130 \times 10^3}{+335} = 388 \text{ K}$

Comment

The decomposition reaction is *feasible* ($\Delta G \leq 0$) above 388 K. Below this temperature, the reverse reaction (the formation of sodium hydrogencarbonate) is feasible. The stoichiometry chosen for the decomposition does not affect the feasibility temperature. Halving the chemical equation (to involve only one mole of $NaHCO_3$) also halves both ΔH and ΔS, leading to the same feasibility temperature.

Essential Notes

Note that the units of ΔH (kilojoules per mole) and ΔS (joules per Kelvin per mole) must be made compatible. This is done by converting kJ mol^{-1} to J mol^{-1} through the factor of 10^3.

Examiners' Notes

The reaction becomes more feasible as the temperature is raised. This finding is in accord with the intuitive belief that sodium hydrogencarbonate is more likely to decompose on heating than on cooling!

The feasibility temperature (388 K) above gives the break-even point for the reaction, i.e. the temperature at which, on balance, the concentrations of reactants and products become roughly equal. In order for the reaction to produce substantially more products than reactants, the temperature will have to be increased somewhat, by some 10 to 20 ${}^{\circ}$C in this case.

The increased temperature moves the equilibrium towards the product side (the *Le Chatelier* effect) and also leads to an increased rate of reaction (the *kinetic* effect); both these features will tend to make the process more viable. Alternatively, if products are removed continuously as they are formed, even quite an unfavourable (unfeasible) process quite distant from its feasibility temperature can yield enough products to become viable.

In the example above, *both ΔH and ΔS are positive*, so the forward reaction is not feasible at low temperatures but becomes *feasible as the temperature is raised*. Other combinations of ΔH and ΔS give different feasibility conditions, as shown in Table 12.

	ΔS positive (entropy gain)	ΔS negative (entropy loss)
ΔH negative (exothermic)	All T	Low T
ΔH positive (endothermic)	High T	No T

Table 12
Favourable temperatures for a feasible reaction under different enthalpy and entropy conditions

3.5.2 Periodicity

The physical properties of the **Period 3 elements** (Na to Ar) were discussed in *Collins Student Support Materials: Unit 1 – Foundation Chemistry*, section 3.1.4. This section considers their chemical reactions.

Study of the reactions of Period 3 elements Na–Ar to illustrate periodic trends

With water

Of the Period 3 elements, only sodium, magnesium and chlorine react with water. The reaction of chlorine with water was studied in *Collins Student Support Materials: Unit 2 – Chemistry in Action*, section 3.2.5. Aluminium, silicon, phosphorus and sulfur do not react with water under normal conditions.

Sodium reacts violently with cold water. A piece of sodium added to cold water fizzes, skates over the surface of the water and becomes molten from the heat of the reaction. Hydrogen is evolved and this may catch fire and burn with a yellow flame (characteristic of sodium). At the end of the reaction, a colourless alkaline solution of sodium hydroxide remains:

$$2Na(s) + 2H_2O(l) \rightarrow 2Na^+(aq) + 2OH^-(aq) + H_2(g)$$

By contrast, as seen in *Collins Student Support Materials: Unit 2 – Chemistry in Action*, section 3.2.6, magnesium reacts only very slowly with cold water but burns in steam when heated to give the white solid magnesium oxide and hydrogen:

$$Mg(s) + H_2O(g) \rightarrow MgO(s) + H_2(g)$$

With oxygen

The solid elements (Na to S) in Period 3 all burn in air or oxygen when ignited. Sodium burns with a yellow flame, forming the oxide:

$$2Na(s) + \tfrac{1}{2}O_2(g) \rightarrow Na_2O(s)$$

Magnesium, aluminium, silicon and phosphorus burn when ignited, emitting a very bright white light and white smoke of the oxides.

$$Mg(s) + \tfrac{1}{2}O_2(g) \rightarrow MgO(s)$$

$$2Al(s) + \tfrac{3}{2}O_2(g) \rightarrow Al_2O_3(s)$$

$$Si(s) + O_2(g) \rightarrow SiO_2(s)$$

$$P_4(s) + 5O_2(g) \rightarrow P_4O_{10}(s)$$

These reactions are very exothermic.

Sulfur burns with a blue flame, but much less vigorously than the elements above, to form the pungent, colourless gas sulfur dioxide:

$$S(s) + O_2(g) \rightarrow SO_2(g)$$

*A survey of the acid–base properties of the oxides
of Period 3 elements*

Physical properties, structure and bonding

The melting points, T_m, of the oxides are summarised in Table 13.

	Na₂O	**MgO**	**Al₂O₃**	**SiO₂**	**P₄O₁₀**	**SO₂**
T_m/K	1548	3125	2345	1883	573	200
Bonding	ionic	ionic	ionic-covalent	covalent	covalent	covalent
Structure	lattice	lattice	lattice	macro-molecular	molecular	molecular

Table 13
Period 3 oxides

Ionic lattices are held together by strong electrostatic forces between ions, so that these lattices have high melting points. Macromolecular solids also have high melting points, because the atoms are held together by strong covalent bonds. Molecular solids involve weak intermolecular dipole–dipole or van der Waals' forces and have low melting points.

Examiners' Notes

Sulfur also forms a higher oxide, SO_3, which can exist in several different forms, all of which have molecular structures with melting points below that of P_4O_{10}.

The reactions of the oxides with water and their structure and bonding are summarised in Table 14.

Reaction with water	pH	Structure and bonding in the oxide
$Na_2O(s) + H_2O(l) \rightarrow 2Na^+(aq) + 2OH^-(aq)$ very soluble; NaOH is a strong alkali	14	ionic lattice
$MgO(s) + H_2O(l) \rightarrow Mg^{2+}(aq) + 2OH^-(aq)$ sparingly soluble; $Mg(OH)_2$ is a weak alkali	9	ionic lattice
no reaction, Al_2O_3 insoluble	7	ionic–covalent lattice
no reaction, SiO_2 insoluble	7	macromolecular covalent
$P_4O_{10} + 6H_2O \rightarrow 4H_3PO_4$ very soluble; violent reaction; H_3PO_4 is a strong acid	0	molecular covalent
$SO_2 + H_2O \rightarrow H_2SO_3$ moderately soluble; H_2SO_3 is a weak acid	3	molecular covalent
$SO_3 + H_2O \rightarrow H_2SO_4$ very soluble; violent reaction; H_2SO_4 is a strong acid	0	molecular covalent

Table 14
Reactions of Period 3 oxides with water and approximate pH values of the resulting solutions

The trend across the period is:

alkaline oxides ⟶ acidic oxides

Across the period, as the bonding in the oxide changes from ionic to molecular, the solutions of the oxides in water change from alkaline to acidic.

Thus the basic oxides of sodium and magnesium react with acids to form salts. The acidic oxides of phosphorus and sulfur, however, react with bases to form salts. In the middle of the period, aluminium oxide being insoluble in water does not affect pH; it is, however, capable of reacting with, and therefore dissolving in, both acids and bases. This behaviour is called **amphoterism** and aluminium oxide is said to be showing **amphoteric character**.

Definition

*An **amphoteric oxide** is one which is capable of reacting with both acids and bases.*

Natural aluminium oxide (e.g. corundum), or laboratory samples which have been heated to red heat, are rather unreactive. Samples prepared at lower temperatures do, however, dissolve in acids to form salts and in alkalis to form aluminates. The amphoteric nature of aluminium hydroxide is discussed in section 3.5.5.

Table 15 gives equations for some examples of these acid–base reactions.

Table 15
Reactions of oxides with acids and bases

Oxide	Reaction with acid/base	
Na_2O	Basic, reacts with acid	$Na_2O + 2HCl \rightarrow 2Na^+ + 2Cl^- + H_2O$
MgO	Basic, reacts with acid	$MgO + 2HCl \rightarrow Mg^{2+} + 2Cl^- + H_2O$
Al_2O_3	Amphoteric, reacts with acid and with base	$Al_2O_3 + 6HCl \rightarrow 2Al^{3+} + 6Cl^- + 3H_2O$ $Al_2O_3 + 2NaOH + 3H_2O \rightarrow 2Na^+ + 2[Al(OH)_4]^-$
SiO_2	Acidic, reacts with base	$SiO_2 + 2NaOH \rightarrow 2Na^+ + SiO_3^{2-} + H_2O$
P_4O_{10}	Acidic, reacts with base	$P_4O_{10} + 12NaOH \rightarrow 12Na^+ + 4PO_4^{3-} + 6H_2O$
SO_2	Acidic, reacts with base	$SO_2 + 2NaOH \rightarrow 2Na^+ + SO_3^{2-} + H_2O$
SO_3	Acidic, reacts with base	$SO_3 + 2NaOH \rightarrow 2Na^+ + SO_4^{2-} + H_2O$

3.5.3 Redox equilibria

Redox equations

In *Collins Student Support Materials: Unit 2 – Chemistry in Action,* section 3.2.4, redox reactions were identified using the **oxidation states** of the elements involved and also the definitions:

> **Definition**
>
> **Oxidation** *is the process of electron loss.*
> **Reduction** *is the process of electron gain.*

These general definitions of oxidation and reduction can also be applied to reactions involving elements and ions in the d block of the Periodic Table.

The ability of **transition elements** to form compounds in which the element is in different oxidation states is one of the most important characteristics of transition elements. This characteristic is central to the behaviour of these elements.

The oxidation state of the transition metal atom in a complex ion is equal to the charge that the element would have if it were a simple ion and not co-ordinated or bonded to other species. The rules which were used to assign oxidation states to elements in s and p blocks of the Periodic Table also apply to **d-block elements**. These rules are given in Table 16.

Examiners' Notes

Transition elements are discussed more fully in section 3.5.4.

Species	Oxidation state
Uncombined elements	0
Combined oxygen, except in peroxides	−2
Combined hydrogen, except in metal hydrides	+1
Combined hydrogen in metal hydrides	−1
Group 1 metals in compounds	+1
Group 2 metals in compounds	+2

Table 16
Rules for assignment of oxidation state

Calculation of oxidation state of d–block elements

Example

Determine the oxidation state of chromium in the complex ion $Cr_2O_7^{2-}$.

As the overall charge is −2, it can be deduced that:
($2 \times$ oxidation state of chromium) + ($7 \times$ oxidation state of oxygen) = −2

Hence:
($2 \times$ oxidation state of chromium) − 14 = −2

Thus:
$2 \times$ oxidation state of chromium = +12

So each chromium in this complex has an oxidation state of +6, i.e. Cr(VI).

Other complex ions

Transition elements form complex ions with a wide variety of different **ligands** (see page 54). When determining the oxidation state of the central transition metal atom in a complex, it is usually far easier to use the overall charge on the ligand rather than the oxidation state of each atom in the ligand. Some common ligands, together with their overall charges, are given in Table 17.

Table 17
Common ligands

Name	Ligand	Overall charge
Water	H_2O	0
Ammonia	NH_3	0
Hydroxide	OH^-	−1
Chloride	Cl^-	−1
Cyanide	CN^-	−1
Ethane-1,2-diamine	$H_2NCH_2CH_2NH_2$	0
Ethanedioate	$C_2O_4{}^{2-}$	−2
Bis[di(carboxymethyl)amino]ethane	$EDTA^{4-}$	−4

The oxidation state of the metal in a complex ion can be worked out by using the same method as shown in the example on page 29; this procedure is demonstrated in the two examples that follow.

Example

Determine the oxidation state of silver in the complex ion $[Ag(CN)_2]^-$

As the overall charge is −1, it can be deduced that:
oxidation state of Ag + (2 × charge on CN^-) = −1

Thus, using the data from Table 17:

oxidation state of Ag − 2 = −1

Hence, Ag has an oxidation state of +1, i.e. Ag(I)

Example

Determine the oxidation state of chromium in the complex ion $[CrCl_2(H_2O)_4]^+$

As the overall charge is +1, it can be deduced that:
oxidation state of Cr + (2 × charge on Cl^-) + (4 × charge on H_2O) = +1
Table 17 shows that Cl has a charge of −1 and that H_2O has zero charge.

Thus:
oxidation state of Cr − 2 = +1

Hence, Cr has an oxidation state of +3, i.e. Cr(III)

The construction of half-equations for reactions

When constructing **half-equations**, the following points must be observed:

- only *one* element in a half-equation changes oxidation state
- the half-equation must balance for atoms
- the half-equation must balance for charge.

For reactions occurring in aqueous solution, it is also helpful to know that, when constructing a half-equation, it can be assumed that water provides a source of oxygen and that any 'surplus' oxygen is converted into water by reaction with hydrogen ions from an acid. Applying these rules, the half-equation for any redox process can be deduced using one of two alternative methods.

Example

Deduce the half-equation for the reduction of VO_3^- to V^{2+} in acid solution.

Method 1

Initial use of oxidation states

In this reaction the oxidation state of vanadium has changed from oxidation state V(V) to oxidation state V(II) and vanadium has been reduced. The number of electrons required for this reduction is deduced first:

$V(V) + 3e^-$ forms $V(II)$

But, as V(V) actually exists as VO_3^- and V(II) exists as the simple ion V^{2+}, an acidic solution is needed to supply six hydrogen ions to combine with three oxygen atoms to form three water molecules. The overall equation is:

$VO_3^- + 6H^+ + 3e^- \rightarrow V^{2+} + 3H_2O$

Finally, the overall charge (+2) on each side of the equation can be used to check that this equation is correct.

Method 2

Initial balancing for atoms

In this reaction VO_3^- changes to V^{2+}.

To balance the equation for atoms, the three oxygen atoms must be combined with six hydrogen ions, provided by added acid, to form three molecules of water:

$VO_3^- + 6H^+$ forms $V^{2+} + 3H_2O$

This equation now balances for atoms but not for charge, with a total charge of +5 on the left-hand side and only +2 on the right-hand side.

For balance, three electrons (i.e. 5 – 2) have to be added to the left-hand side to give the half-equation:

$$VO_3^- + 6H^+ + 3e^- \rightarrow V^{2+} + 3H_2O$$

Finally, the change in oxidation state of vanadium can be used to check that this equation is correct.

The construction of overall equations for redox reactions

The overall equation for any redox reaction can be obtained by adding together two half-equations, so that the number of electrons given by the reducing agent exactly balances the number of electrons accepted by the oxidising agent.

If the same species appears on both sides of an overall equation obtained by adding two half-equations, this species must be cancelled to give a simpler equation. Water molecules and hydrogen ions are the most common species which need to be treated in this way.

Example

In acidic solution, dichromate(VI) ions, $Cr_2O_7^{2-}$, can be reduced to chromium(III), Cr^{3+}, by sulfite ions, SO_3^{2-}, which are oxidised to sulfate ions, SO_4^{2-}. Derive half-equations for the oxidation of sulfite ions to sulfate ions and for the reduction of dichromate(VI) ions to chromium(III). Use these half-equations to derive an equation for the overall reaction.

In the half-equation for the reduction of chromium from oxidation state +6 to +3, an acidic solution is required and the 'surplus' oxygen forms water:

$$Cr_2O_7^{2-} + 14H^+ + 6e^- \rightarrow 2Cr^{3+} + 7H_2O$$

In the half-equation for the oxidation of sulfur from oxidation state +4 to oxidation state +6, the additional oxygen has been 'supplied' by a water molecule:

$$SO_3^{2-} + H_2O \rightarrow SO_4^{2-} + 2H^+ + 2e^-$$

The first half-equation shows that six electrons must be supplied for each dichromate(VI) ion reduced. Since the oxidation of each sulfite ion provides only two electrons, then three of these oxidation half-reactions are needed for each dichromate(VI) ion reduction half-reaction. The overall equation for the redox reaction is obtained by addition:

$$Cr_2O_7^{2-} + 14H^+ + 6e^- \rightarrow 2Cr^{3+} + 7H_2O$$

$$3SO_3^{2-} + 3H_2O \rightarrow 3SO_4^{2-} + 6H^+ + 6e^-$$

$$\overline{Cr_2O_7^{2-} + 14H^+ + 3SO_3^{2-} + 3H_2O \rightarrow 2Cr^{3+} + 7H_2O + 3SO_4^{2-} + 6H^+}$$

Essential Notes

When adding two half-equations together, the electrons **must** cancel. Other species, such as H^+ or H_2O, may cancel.

In this example it is necessary to simplify the equation by cancelling three water molecules and six hydrogen ions to give the overall equation:

$$Cr_2O_7{}^{2-} + 8H^+ + 3SO_3{}^{2-} \rightarrow 2Cr^{3+} + 4H_2O + 3SO_4{}^{2-}$$

Electrode potentials

Half-equations, electron transfer, reduction and oxidation

Half-equations (which represent half-reactions) have already been discussed in *Collins Student Support Materials: Unit 2 – Chemistry in Action*, section 3.2.4, and also in the 'Redox equations' section (pages 29–32). Electron transfer leads to electron gain by one species and electron loss by another. Electron gain is called **reduction**, electron loss is called **oxidation**. All redox reactions can be expressed as the sum of two half-reactions – one involving electron gain, and the other electron loss.

The IUPAC convention for writing half-equations

By convention, *all* redox half-equations are written as *reductions* (electron addition). This is the convention of the *International Union of Pure and Applied Chemistry* (IUPAC). Two such half-equations, written conventionally, are shown below:

$$Fe^{3+}(aq) + 3e^- \rightarrow Fe(s)$$
$$\tfrac{1}{2}Br_2(l) + e^- \rightarrow Br^-(aq)$$

Electrochemical cells

Redox reactions can be studied electrically using an **electrochemical cell**. A cell contains two **electrodes** (electrical conductors) immersed in an **electrolyte** (an ionic conductor), either as an aqueous solution or as a molten salt. An electrode together with its associated electrolyte forms an **electrode compartment**. Sometimes both electrodes share the same compartment (Fig 2) but if two compartments with different electrolytes are used, they are connected by means of a **salt bridge** (Fig 3).

A salt bridge consists of an electrolyte solution (often a saturated solution of KCl or KNO_3 in agar jelly), which completes the electrical circuit. It enables the cell to work by allowing **ions** to move between the two compartments while keeping apart the different solutions in the two compartments. If a wire were to be used instead of a salt bridge, it would introduce two more electrodes into the circuit.

Electrode reactions

In an electrochemical cell, each electrode compartment supports its own half-reaction. Electrons released by the oxidation half-reaction in one compartment, for example:

$$Zn(s) \rightarrow Zn^{2+}(aq) + 2e^-$$

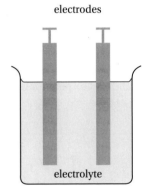

Fig 2
A cell with a shared common electrolyte and a single compartment

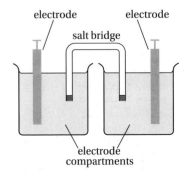

Fig 3
A cell with two electrolytes in separate compartments connected by a salt bridge

are made available to drive the reduction half-reaction in the other compartment, for example:

$$Cu^{2+}(aq) + 2e^- \rightarrow Cu(s)$$

by flowing through the external circuit that completes electrical contact between the two compartments.

Electrodes

The simplest of all is the **metal electrode**, which comprises a metal in equilibrium with a solution of its ions. An example is the copper electrode which, depending on the position it occupies in the cell, is written as:

$$Cu(s)\,|\,Cu^{2+}(aq) \; or \; Cu^{2+}(aq)\,|\,Cu(s)$$

and is called the electrode representation. The vertical bars ($|$) shown above denote boundaries between different phases, which are themselves specified using state symbols.

The redox couple (e.g. Cu^{2+}/Cu) is a shorthand way of writing a reduction half-equation (e.g. $Cu^{2+}(aq) + 2e^- \rightarrow Cu(s)$). By convention, the oxidised species (without a state symbol) is written first, separated from the reduced species (also without a state symbol) by a forward slash (i.e. *Ox/Red*). This redox couple notation is not used in *cell diagrams* (see page 35) but is an economical way of specifying the half-equation in question.

A **gas electrode** consists of an inert metal (usually platinum) surrounded by a gas in equilibrium with a solution of its ions. The inert metal simply acts as either a source or a sink for electrons. The **hydrogen electrode**, whose special significance will be discussed shortly, is shown schematically in Fig 4 on page 36.

The hydrogen electrode corresponds to the redox couple H^+/H_2 and the electrode is denoted conventionally as:

Left electrode $Pt(s)\,|\,H_2(g)\,|\,H^+(aq) \; or \; H^+(aq)\,|\,H_2(g)\,|\,Pt(s)$ *Right electrode*

with *two* phase boundaries (solid–gas and gas–liquid) shown as vertical bars. The reduction half-reaction at this electrode is:

$$2H^+(aq) + 2e^- \rightarrow H_2(g)$$

A **redox electrode** is one at which two oxidation states of a given element undergo a reduction reaction at an inert metal surface, as in the case of the Fe^{3+}/Fe^{2+} couple:

Left $Pt(s)\,|\,Fe^{2+}(aq), Fe^{3+}(aq) \; or \; Fe^{3+}(aq), Fe^{2+}(aq)\,|\,Pt(s)$ *Right*

which has a reduction half-reaction, and a corresponding half-equation, written as:

$$Fe^{3+}(aq) + e^- \rightarrow Fe^{2+}(aq)$$

The redox couple $MnO_4^-, H^+/Mn^{2+}$ would be denoted as an electrode by:

Left $Pt(s)\,|\,Mn^{2+}(aq), MnO_4^-(aq), H^+(aq) \; or$
$MnO_4^-(aq), H^+(aq), Mn^{2+}(aq)\,|\,Pt(s)$ *Right*

corresponding to the reduction half-equation:

$$MnO_4^-(aq) + 8H^+(aq) + 5e^- \rightarrow Mn^{2+}(aq) + 4H_2O(l)$$

Essential Notes

Shorthand symbols like Cu^{2+}/Cu or Fe^{3+}/Fe are known as **redox couples**.
The standard electrode potentials of the reduction equations that they represent: e.g. $Cu^{2+}(aq) + 2e^- \rightarrow Cu(s)$ and $Fe^{3+}(aq) + 3e^- \rightarrow Fe(s)$ are called **redox potentials**.

Examiners' Notes

By convention, cell diagrams (see page 35) are always written with metal electrodes on the *outside*. For the electrode on the right, the electrode components are written in the same order as the couple (*Ox/Red*) with the reverse order (*Red/Ox*) used for the electrode on the left. Reading through a cell from left to right the order is *Red/Ox/Ox/Red* or **ROOR**.

Combining half-reactions into cell diagrams, and writing cell equations

A **cell diagram** is constructed by writing two electrodes back to back, and joining them with a salt bridge, conventionally denoted by two vertical bars.

$$Zn(s)\,|\,Zn^{2+}(aq)\,\|\,Cu^{2+}(aq)\,|\,Cu(s)$$

There is a simple convention which helps when drawing a cell diagram with two electrode compartments joined together to form a cell:

Cell convention

The cell diagram is written with the more positive electrode (the one at which reduction occurs) shown as the right-hand electrode.

Thus, in the zinc/copper cell above, the more positive electrode (Cu/Cu^{2+}) is correctly placed on the right-hand side of the cell diagram.

The **cell reaction** that corresponds to this cell diagram can be derived in one of two entirely equivalent ways, as follows:

Subtraction Method	Addition Method
• write the right-hand half-reaction as a *reduction*	• write the right-hand half-reaction as a *reduction*
• beneath it, write the left-hand half-reaction, also as a *reduction*	• beneath it, reverse the left-hand half-reaction, and write it as an *oxidation*
• *subtract* the left-hand (reduction) half-reaction from the right-hand one	• *add* the left-hand half-reaction (oxidation) to the right-hand one (reduction)
Right (red): $Cu^{2+}(aq) + 2e^- \rightarrow Cu(s)$	Right (red): $Cu^{2+}(aq) + 2e^- \rightarrow Cu(s)$
Left (red): $Zn^{2+}(aq) + 2e^- \rightarrow Zn(s)$	Left (ox): $Zn(s) \rightarrow Zn^{2+}(aq) + 2e^-$
Overall (by subtraction): $Cu^{2+}(aq) + Zn(s) \rightarrow Cu(s) + Zn^{2+}(aq)$	Overall (by addition): $Cu^{2+}(aq) + Zn(s) \rightarrow Cu(s) + Zn^{2+}(aq)$

This is the way the cell reaction actually goes; zinc does displace copper from solution.

Cell reactions

With the more positive electrode (the one at which reduction occurs) shown as the right-hand electrode, the cell reaction goes in the forward direction, as written.

Essential Notes

The double bar signifies a junction across which there is no potential difference – usually a **salt bridge**. Sometimes it is shown as two vertical dashed bars.

Examiners' Notes

As is self-evident, these two methods are entirely equivalent. Which one is chosen is simply a matter of personal preference.

Cell potential

In an electrochemical cell, a potential difference (*voltage*) is set up between the two electrodes. By convention, the left-hand electrode is more negative, relative to the right-hand electrode, which is more positive. Each electrode takes up its own characteristic potential and the overall **cell potential** is the difference between the two.

Factors that affect the cell potential

- *Cell current* – the true cell potential can be measured only under *zero-current* conditions, most commonly using a high-resistance digital voltmeter. The cell potential measured with zero current is called the *electromotive force* or the *e.m.f.* (see below).

- *Cell concentration* – solution concentrations affect the cell potential. The *standard concentration* chosen is 1.00 mol dm^{-3}.

- *Cell temperature* – temperature affects the cell potential. The *standard temperature* chosen is 298 K.

- *Cell pressure* – pressure affects cell potentials, but not significantly unless a gas electrode is used. The *standard pressure* chosen is 100 kPa (1 bar) exactly.

The standard electromotive force (e.m.f.)

Under standard conditions (zero current, 1.00 mol dm^{-3}, 298 K, 100 kPa) the cell potential is known as the *standard e.m.f. of the cell*. The **cell e.m.f.** does not have a special symbol.

Examiners' Notes

If current is being drawn from the cell, the potential drops as the concentrations of the solutions in the cell change.

Definition

*The **standard electromotive force (e.m.f.)** is the potential difference between the electrodes of a standard electrochemical cell measured under zero-current conditions.*

Essential Notes

The SI unit of electric potential difference is the volt (V).

Standard Hydrogen Electrode (SHE)

It is impossible to measure the potential of a single electrode on its own, but if one electrode is assigned the value *zero*, then all other electrode potentials can be listed relative to this standard. The standard chosen is the hydrogen electrode operating under standard conditions.

It is called the **standard hydrogen electrode** (SHE).

$$Pt(s) \,|\, H_2(g,\ 100\ kPa) \,|\, H^+(aq,\ 1.00\ mol\ dm^{-3}) \qquad E^{\ominus} = 0 \text{ at } 298\ K$$

The basic form of this electrode was described earlier (see gas electrode on page 34) and is shown schematically in Fig 4. Under standard conditions, $E^{\ominus}$(SHE) = 0, by definition.

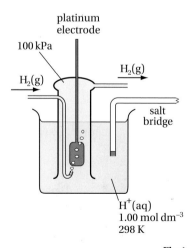

platinum electrode

100 kPa

$H_2(g)$

$H_2(g)$

salt bridge

$H^+(aq)$
1.00 mol dm^{-3}
298 K

Fig 4
The hydrogen electrode. Hydrogen gas is bubbled over a platinum electrode establishing an equilibrium with $H^+(aq)$

Definition

*The potential, $E^{\ominus}$(SHE), of the **standard hydrogen electrode** is zero.*

Standard electrode potential

The **standard electrode potential**, of any redox system or couple is found by measuring the potential of a cell with the SHE as the left-hand electrode and the unknown as the right-hand one.

All the variables affecting cell potential listed above *must* be kept under standard conditions when determining standard potentials.

Calculating $E^\ominus$ for an electrode from the *cell e.m.f.*

The overall cell reaction is determined by subtracting the left-hand half-equation from the right-hand half-equation. The e.m.f. is determined by subtracting the left-hand (L) electrode potential from the right-hand (R) one. Thus:

$$cell\ e.m.f. = E^\ominus(R) - E^\ominus(L)$$

If the left-hand electrode is the SHE with $E^\ominus = 0$, then $E^\ominus_{right} = cell\ e.m.f.$

To ensure that the cell reaction goes in the forward direction, the less positive standard potential is put to the left and is subtracted from the right-hand one, giving a *cell e.m.f.* > 0.

Some illustrations of the uses of electrode potentials and of *cell e.m.f.* are shown in the following examples.

Example

Calculate the standard e.m.f. of a cell with a silver electrode (Ag^+/Ag, $E^\ominus = +0.80\ V$) and a copper electrode (Cu^{2+}/Cu, $E^\ominus = +0.34\ V$), each in its own solution of 1.0 mol dm^{-3} silver(I) or 1.0 mol dm^{-3} copper(II) ions connected by a salt bridge.

Method

The more positive potential (Ag^+/Ag) is made the *right-hand electrode.*

Answer

$$cell\ e.m.f. = E^\ominus_{right} - E^\ominus_{left}$$

Therefore *cell e.m.f.* = +0.80 − (+0.34) = +0.46 V

Example

Calculate the standard electrode potential of an unknown metal M which, in a standard solution of its ions $M^{n+}(aq)$, gives a *cell e.m.f.* of +0.48 V when acting as the right-hand electrode in a standard cell with tin ($E^\ominus$ (Sn^{2+}/Sn) = −0.14 V) as the left-hand electrode.

Method

Use:

$$cell\ e.m.f. = E^\ominus_{right} - E^\ominus_{left}$$

Answer

M^{n+}/M is the right-hand electrode and $E^{\ominus}_{\text{right}} = cell\ e.m.f. + E^{\ominus}_{\text{left}}$

$E^{\ominus}(M^{n+}/M) = +0.48 + (-0.14) = +0.34\ V$

Example

Calculate the e.m.f. of a cell with the *secondary standard* silver/silver chloride electrode ($E^{\ominus} = +0.22\ V$) as the left-hand electrode and the Fe^{2+}/Fe electrode ($E^{\ominus} = -0.44\ V$) as the right-hand one.

Method

Use:

$$cell\ e.m.f. = E^{\ominus}_{\text{right}} - E^{\ominus}_{\text{left}}$$

Answer

cell e.m.f. $= -0.44 - 0.22 = -0.66\ V$

Comment

The *cell e.m.f.* is negative, so the cell reaction will *not* go in the forward direction. Instead, the reverse reaction occurs:

$$2AgCl(s) + Fe(s) \rightarrow 2Ag(s) + Fe^{2+}(aq) + 2Cl^-(aq)$$

Electrochemical series

Standard electrode potentials measured relative to the SHE can be placed in a list, either descending in voltage from the most positive (the convention adopted here) or ascending from the most negative. The choice is arbitrary, and it is necessary to accept and understand data presented in either format.

The data in Table 18 forms a part of the **electrochemical series**, with reduction half-reactions listed in order of decreasing electrode potential. In each case, the reactions listed can be thought of as an oxidising agent (on the left-hand side of a reduction half-equation) accepting electrons to form a reducing agent (on the right-hand side of a reduction half-equation).

- The strongest *oxidising* agents accept electrons easily and have more positive potentials.

- The strongest *reducing* agents lose electrons easily and have more negative potentials.

- Half-reactions with more positive potentials correspond to electron gain (**reduction**) reactions and go readily from left to right.

- Half-reactions with more negative potentials correspond to electron loss (**oxidation**) reactions and go readily from right to left.

Examiners' Notes

Because the SHE is complicated to set up and to use, a more convenient secondary standard electrode, already calibrated against the SHE, is used instead. So, an equally valid determination of an unknown standard potential may be made by measuring relative to a **secondary standard** electrode. Knowledge of secondary cells is not required in the AQA A-level specification.

Examiners' Notes

Many of the helpful concepts for interpreting and using the electrochemical series, including those used in this book, depend on the order in which the data are presented. Flexibility of approach is needed to cope with any order of presentation.

Reduction half-reaction ox + e⁻ → red	$E^{\ominus}$/V
$F_2(g) + 2e^- \rightarrow 2F^-(aq)$	+2.87
$MnO_4^{2-}(aq) + 4H^+(aq) + 2e^- \rightarrow MnO_2(s) + 2H_2O(l)$	+1.55
$MnO_4^-(aq) + 8H^+(aq) + 5e^- \rightarrow Mn^{2+}(aq) + 4H_2O(l)$	+1.51
$Cl_2(g) + 2e^- \rightarrow 2Cl^-(aq)$	+1.36
$Cr_2O_7^{2-}(aq) + 14H^+(aq) + 6e^- \rightarrow 2Cr^{3+}(aq) + 7H_2O(l)$	+1.33
$Br_2(l) + 2e^- \rightarrow 2Br^-(aq)$	+1.09
$Ag^+(aq) + e^- \rightarrow Ag(s)$	+0.80
$Fe^{3+}(aq) + e^- \rightarrow Fe^{2+}(aq)$	+0.77
$MnO_4^-(aq) + e^- \rightarrow MnO_4^{2-}(aq)$	+0.56
$I_2(s) + 2e^- \rightarrow 2I^-(aq)$	+0.54
$Cu^{2+}(aq) + 2e^- \rightarrow Cu(s)$	+0.34
$Hg_2Cl_2(aq) + 2e^- \rightarrow 2Hg(l) + 2Cl^-(aq)$	+0.27
$AgCl(s) + e^- \rightarrow Ag(s) + Cl^-(aq)$	+0.22
$2H^+(aq) + 2e^- \rightarrow H_2(g)$	defined as 0
$Pb^{2+}(aq) + 2e^- \rightarrow Pb(s)$	−0.13
$Sn^{2+}(aq) + 2e^- \rightarrow Sn(s)$	−0.14
$V^{3+}(aq) + e^- \rightarrow V^{2+}(aq)$	−0.26
$Fe^{2+}(aq) + 2e^- \rightarrow Fe(s)$	−0.44
$Zn^{2+}(aq) + 2e^- \rightarrow Zn(s)$	−0.76
$Al^{3+}(aq) + 3e^- \rightarrow Al(s)$	−1.66
$Mg^{2+}(aq) + 2e^- \rightarrow Mg(s)$	−2.36
$Na^+(aq) + e^- \rightarrow Na(s)$	−2.71
$Ca^{2+}(aq) + 2e^- \rightarrow Ca(s)$	−2.87
$K^+(aq) + e^- \rightarrow K(s)$	−2.93
$Li^+(aq) + e^- \rightarrow Li(s)$	−3.05

Increasing Oxidising Power (arrow pointing up, left side)

Increasing Reducing Power (arrow pointing down)

Table 18
Standard electrode potentials at 298 K

The use of $E^{\ominus}$ values from the Electrochemical Series to predict the direction of simple redox reactions

The cell reaction occurs in the forwards direction of the cell equation if the right-hand electrode is the site of *reduction*. When this happens, the right-hand electrode is more positive than the left–hand electrode and $E^{\ominus}_{\text{right}} - E^{\ominus}_{\text{left}}$ is necessarily positive.

The direction of redox reactions

A cell reaction goes forwards, in the direction written, if, and only if, the corresponding *cell e.m.f.* is positive.

In terms of the electrochemical series, a half-reaction with a more positive potential *oxidises* one with a more negative potential. In terms of the reduction equations in Table 18, the spontaneous direction of reaction (left to right, or right to left) involving pairs of half-reactions can be summarised as:

Cell e.m.f.	Spontaneous direction
+ve	Forwards
–ve	Backwards

Essential Notes

Reminder:

Shorthand symbols like Zn^{2+}/Zn are known as *redox couples*.

The standard electrode potentials of the reduction equations that they represent:

e.g. $Zn^{2+}(aq) + 2e^- \rightarrow Zn(s)$ are called *redox potentials*.

The electrochemical series and *cell e.m.f.*

Predictions about redox reactions can be made using the half-reactions in a table of standard redox potentials. Below are shown two methods of approach. Each of these methods is actually equivalent to the other, though at first sight they do appear to be different.

(i) Method 1 (the 'six-step' approach):

A series of six simple steps will always lead to the right answer. These are illustrated in Table 19, using the Ag^+/Ag and the Zn^{2+}/Zn redox couples as an example.

Table 19
Six steps in calculating *cell e.m.f.*

Step 1

Consider the two half-reactions:

$$Ag^+(aq) + e^- \rightarrow Ag(s) \qquad E^\ominus = +0.80\,V$$
$$Zn^{2+}(aq) + 2e^- \rightarrow Zn(s) \qquad E^\ominus = -0.76\,V$$

Step 2

The half-equation with the *more positive* $E^\ominus$ value becomes the *positive electrode*; electrons arrive here from the external circuit, so:

$$Ag^+(aq) + e^- \rightarrow Ag(s)$$

The conventional half-reaction goes *forwards*.
Silver ions behave as the *oxidising* agent and are reduced.

Step 3

The half-equation with the *less positive* $E^\ominus$ value becomes the *negative electrode*; electrons leave this electrode and enter the external circuit, so:

$$Zn(s) \rightarrow Zn^{2+}(aq) + 2e^-$$

The conventional half-reaction goes *backwards*.
Zinc atoms behave as the *reducing* agent and are oxidised.

Step 4

The *overall equation* is obtained by adding the two new half-reactions (step 2 and step 3) so that *electrons cancel*:

$$2Ag^+(aq) + Zn(s) \rightarrow Ag(s) + Zn^{2+}(aq)$$

Note the need to double the Ag^+/Ag equation so that the electrons can cancel.

Step 5

The *cell representation* is obtained by placing the couple with the *more positive* $E^\ominus$ value on the right, with a salt bridge to separate it from the other couple:

$$\overset{\ominus}{}Zn(s)\,|\,Zn^{2+}(aq)\,||\,Ag^+(aq)\,|\,Ag(s)\overset{\oplus}{}$$

The electrode polarities follow the rule: *positive on the right*.

Step 6

The **cell e.m.f.** is obtained using *cell e.m.f.* $= E^\ominus_{right} - E^\ominus_{left}$, so:

$$cell\ e.m.f. = +0.80 - (-0.76) = +1.56\,V$$

(ii) Method 2 (the 'outline sketch' approach):

Several of the stages listed in Table 19, particularly steps 1, 2, 3 and 5, can be represented as a diagram. This procedure is shown in Fig 5 and Fig 6 for the reaction between the Ag^+/Ag and the Zn^{2+}/Zn couples. A sketch based on the electrochemical series (Table 18) is drawn and the two selected half-reactions are identified (*step 1*, Table 19). The *higher* (more positive) couple is marked ⊕ and the *lower* (less positive) one is marked ⊖ (*steps 2 and 3*, Table 19). All other (intervening and outlying) half-equations and potentials in the electrochemical series can be ignored.

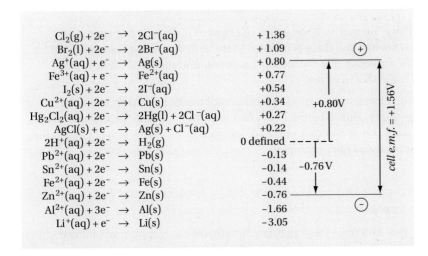

Fig 5
Determining the direction of spontaneous reaction and e.m.f. for a redox couple

The graphical procedure is simplified by sketching the potential horizontally, so that it increases from left to right rather than upwards. Using the two chosen half-equations only, the diagram in Fig 5 is turned through 90° clockwise to fit the rule *'positive potential to the right'* (*step 5*, Table 19) and the reaction half-equations are omitted.

The result is an **outline sketch**, as in Fig 6, from which the cell representation can be written by simple inspection (*step 5*, Table 19). The direction of the cell reaction and the *cell e.m.f.* (*steps 4 and 6*, Table 19) can then be determined using the rule *'R – L'*, or by reversing the oxidation (left-hand) equation and then adding.

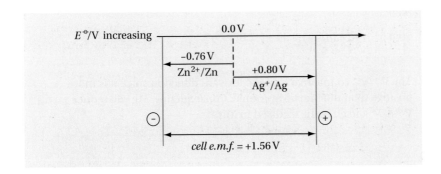

Fig 6
Outline sketch to determine the direction of spontaneous reaction and e.m.f. for a redox couple

In the example shown, the cell representation is:

$$\overset{\ominus}{}Zn(s)\,|\,Zn^{2+}(aq)\,||\,Ag^{+}(aq)\,|\,Ag(s)\overset{\oplus}{}$$

the *spontaneous cell reaction* is:

Positive electrode:	$2Ag^{+}(aq) + 2e^{-} \rightarrow 2Ag(s)$
Negative electrode:	$Zn^{2+}(aq) + 2e^{-} \rightarrow Zn(s)$
Overall cell reaction (R – L):	$2Ag^{+}(aq) + Zn(s) \rightarrow Zn^{2+}(aq) + 2Ag(s)$

and the *cell e.m.f.* is:

$$cell\ e.m.f. = E_{right}^{\ominus} - E_{left}^{\ominus} = +0.80 - (-0.76) = +1.56\,V$$

These alternative approaches (*six steps* or *outline sketch*) are entirely equivalent and there is no need to try both. Instead, it is worth choosing the approach that comes most naturally and then practising to gain confidence in its use.

The determination of e.m.f. and overall reaction is illustrated in the examples shown below. In the first example, both the *'six step'* and the *'basic sketch'* methods are used, followed by an illustration of each in the second and third examples.

Example

Use data from Table 18 to predict whether magnesium will reduce vanadium(III) ions to vanadium(II) ions. Write the representation of a standard cell in which reaction would occur and determine its e.m.f.

Method 1

Carry out the *six steps* in Table 19:

Step 1: $Mg^{2+}(g) + 2e^{-} \rightarrow Mg(s)$ $E^{\ominus} = -2.73\,V$
          $V^{3+}(aq) + 2e^{-} \rightarrow V^{2+}(aq)$ $E^{\ominus} = -0.26\,V$

Step 2: $V^{3+}(aq) + 2e^{-} \rightarrow V^{2+}(aq)$ $V^{3+}(aq)$ is the *oxidising* agent

Step 3: $Mg(s) \rightarrow Mg^{2+}(aq) + 2e^{-}$ $Mg(s)$ is *oxidised*

Step 4: $2V^{3+}(aq) + Mg(s) \rightarrow Mg^{2+}(aq) + 2V^{2+}(aq)$

Step 5: $\overset{\ominus}{}Mg(s)\,|\,Mg^{2+}(aq)\,||\,V^{3+}(aq),V^{2+}(aq)\,|\,Pt(s)\overset{\oplus}{}$

Step 6:   *cell e.m.f.* $= -0.26 - (-2.37) = +2.11\,V$

Comment

The V^{3+}/V^{2+} couple reacts in the forward direction since it is more positive than the Mg^{2+}/Mg couple. Consequently, Mg will reduce V^{3+} to V^{2+}, itself being oxidised to Mg^{2+}.

Method 2

Draw the *outline sketch* as in Fig 6 on page 41:

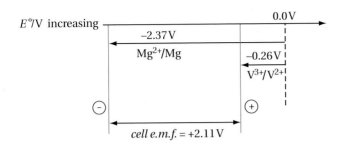

By inspection,

cell representation:

$\ominus$Mg(s) | Mg^{2+}(aq) || V^{3+}(aq), V^{2+}(aq) | Pt(s)$\oplus$

cell reaction (R – L):

Mg(s) + 2V^{3+}(aq) $\rightarrow$ Mg^{2+}(aq) + 2V^{2+}(aq)

cell e.m.f. (R – L):

cell e.m.f. = –0.26 – (–2.37) = +2.11 V

Comment

The V^{3+}/V^{2+}couple reacts in the forward direction since it is more positive than the Mg^{2+}/Mg couple. Consequently, Mg will reduce V^{3+} to V^{2+}, itself being oxidised to Mg^{2+}.

Example

Use data from Table 18 to determine whether chlorine can oxidise Fe(II) ions to Fe(III) ions. Write the representation of a standard cell in which this reaction might occur and determine its e.m.f.

Method 1

Carry out the six steps in Table 19.

Step 1:   Cl$_2$(g) + 2e$^-$ $\rightarrow$ 2Cl$^-$(aq) $E^{\ominus}$ = +1.36 V

          Fe^{3+}(aq) + e$^-$ $\rightarrow$ Fe^{2+}(aq) $E^{\ominus}$ = +0.77 V

Step 2:   Cl$_2$ + 2e$^-$ $\rightarrow$ 2Cl$^-$ Cl$_2$ is the *oxidising* agent

Step 3:   Fe^{2+} $\rightarrow$ Fe^{3+} + e$^-$ Fe^{2+} is *oxidised*

Step 4:   2Fe^{2+}(aq) + Cl$_2$(g) $\rightarrow$ 2Fe^{3+}(aq) + 2Cl$^-$(aq)

Step 5: $\ominus$Pt(s) | Fe^{2+}(aq), Fe^{3+}(aq) || Cl$^-$(aq) | Cl$_2$(g) | Pt(s)$\oplus$

Step 6:   *cell e.m.f.* = +1.36 – (+0.77) = +0.59 V

Comment

The Cl_2/Cl^- couple reacts in the forward direction because it is more positive than the Fe^{3+}/Fe^{2+} couple. Consequently, Cl_2 will oxidise Fe^{2+} to Fe^{3+}, itself being reduced to Cl^-.

Example

Can manganate(VI) ions in aqueous solution disproportionate into manganese(VII) and manganese(IV) species? Write the representation of a standard cell in which this reaction might occur and determine its e.m.f. Use data from Table 18.

Method 2

Draw the basic sketch as in Fig 6 on page 41:

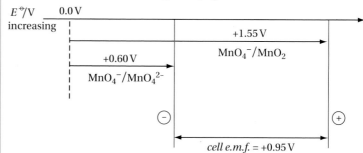

By inspection,

cell representation:

$$^{\ominus}Pt(s)\,|\,MnO_4^{2-}(aq),\ MnO_4^-(aq)\,||\,MnO_4^{2-}(aq)\,|\,MnO_2(s)\,|\,Pt(s)^{\oplus}$$

cell reaction (R – L):

$$4H^+(aq) + 3MnO_4^{2-}(aq) \rightarrow 2MnO_4^-(aq) + MnO_2(s) + 2H_2O(l)$$

cell e.m.f. (R – L):

$$cell\ e.m.f. = +1.55 - (+0.60) = +0.95\ V$$

Comment

The MnO_4^{2-}/MnO_2 couple reacts as shown in the equation (i.e. in the forward direction) because it is more positive than the MnO_4^-/MnO_4^{2-} couple, so manganate(VI) ions in aqueous solution can disproportionate into manganese(VII) and manganese(IV) species.

The effect of conditions on the *cell e.m.f.*
It is important to have a set of defined conditions when measuring standard electrode potentials, as the conditions chosen can change the *cell e.m.f.*

The effects of changing cell conditions can be predicted using Le Chatelier's principle. In this case, however, there is no direct chemical equilibrium between the two half-cells. Instead, the response of the cell to changed conditions arises entirely through a change in the e.m.f., upwards or downwards, in response to any constraint applied to the cell.

The e.m.f. is a measure of how far from chemical equilibrium the cell reaction actually lies; the more positive the e.m.f., the more likely the forward cell reaction becomes. If a cell is allowed to run down, then the cell potential drops to zero.

Similar arguments apply to the other cell variables:

- *cell current* – once current starts to be drawn from a standard cell the e.m.f. *drops*; it reaches *zero* once chemical equilibrium has been established.

- *cell concentration* – it is quite straightforward to apply Le Chatelier's principle to a cell reaction, as can be seen in the example of the Daniel cell equation, which is written below as an equilibrium in order to emphasise the link with Le Chatelier's principle:

$$Cu^{2+}(aq) + Zn(s) \rightleftharpoons Cu(s) + Zn^{2+}(aq)$$

The equilibrium in this reaction lies well over to the right (lots of Zn^{2+}, not much Cu^{2+}) with the e.m.f. dropping to zero as chemical equilibrium is approached.

Applying Le Chatelier's principle, increasing the concentration of copper(II) ions will push the equilibrium to the right. But, as the two half-cells are joined electrically rather than chemically, the concentrations cannot adjust in this way. Instead, the *cell e.m.f.* alters to accommodate the change in the concentration of Cu^{2+}. Concentration changes away from equilibrium (more Cu^{2+}, less Zn^{2+}) cause the e.m.f. to *increase* and concentration changes towards equilibrium (less Cu^{2+}, more Zn^{2+}) cause the e.m.f. to *decrease*.

- *cell temperature* – cell reactions with positive e.m.f. values are driven by favourable enthalpies of reaction; they are often very exothermic and are enthalpy driven, with entropy playing only a secondary role. Hence, again applying Le Chatelier's principle, an *increase* in temperature will favour the backward reaction, thereby *decreasing* the e.m.f.

These effects are summarised in Table 20 for a typical redox cell:

$$Pt(s) \mid V^{2+}(aq), V^{3+}(aq) \parallel Fe^{3+}(aq), Fe^{2+}(aq) \mid Pt(s) \qquad E^{\ominus}/V = +1.03$$

with half-reactions:

Right: $Fe^{3+}(aq) + e^- \rightarrow Fe^{2+}(aq)$

Left: $V^{3+}(aq) + e^- \rightarrow V^{2+}(aq)$

and cell reaction:

$$Fe^{3+}(aq) + V^{2+}(aq) \rightleftharpoons V^{3+}(aq) + Fe^{2+}(aq) \qquad \Delta H < 0$$

Essential Notes

As the cell reaction approaches equilibrium, so the e.m.f. drops towards zero.

Examiners' Notes

The cell potential is well named; it denotes the potential of the cell to do useful work by running the cell equation from left to right.

For this reason, cell reactions are not usually written with equilibrium arrows, as they are not in chemical equilibrium, but will, if the compartments are mixed, proceed spontaneously from left to right.

Examiners' Notes

Conditions that favour the *forward* reaction result in an *increase* in the e.m.f.; those that favour the *backward* reaction result in a *decrease* in the e.m.f.

In general terms, any such reaction could be written as:

$$R_{ox} + L_{red} \rightleftharpoons L_{ox} + R_{red}$$

where R and L refer to right-hand and left-hand cells respectively, with the subscripts denoting the oxidised and reduced species in each half-cell.

Table 20
The effects of conditions on *cell e.m.f.*

	Increase in				
	$R_{ox}=[Fe^{3+}]$	$L_{red}=[V^{2+}]$	$L_{ox}=[V^{3+}]$	$R_{red}=[Fe^{2+}]$	Temperature
Effect on *cell e.m.f.*	increase	increase	decrease	decrease	decrease

Examiners' Notes

Altering cell concentrations or the temperature affects both $E^{\ominus}$ and e.m.f. values, but the resulting changes are relatively small.

Changes in conditions have predictable effects on the *cell e.m.f.* However, the changes in e.m.f. caused by moderate changes in concentration or temperature are rarely very large. For example, an increase in temperature of 10 °C in the iron/vanadium cell above will lower the e.m.f. from 1.03 V to 1.01 V.

Electrochemical cells

Specially adapted electrochemical cells are commonly used as convenient sources of electrical energy. Such cells can be divided into three main categories: **primary cells, secondary cells** and **fuel cells**:

Examiners' Notes

Fuel cells can also be called *primary cells*, in that they are charged by the oxidation of a continuous supply of fuel and are **not** recharged by an electric current.

Definition

A *primary cell* is *irreversible and is not intended to be recharged by an electric current.*

A *secondary cell* is *reversible and is specifically designed to be recharged by an electric current.*

A *fuel cell* generates electricity from the continuous oxidation of an external source of fuel.

Electrochemical cells that are used to provide a convenient source of electrical power are commonly called 'batteries', though the term 'battery' originates from an array of more than one cell joined in series to others.

Examiners' Notes

Knowledge of the details of these cells is **not** required in the AQA A-level specification; all that is needed is an ability to manipulate given equations and potentials to yield e.m.f. values or vice-versa.

Practical cells

Primary cells

Primary cells are *irreversible cells*. The single function of a primary electrical cell is to provide current to an external circuit while discharging (galvanic action). Once discharged, primary cells are discarded.

Characteristic features of some common primary cells are shown in Table 21.

Table 21

Primary cells (non-rechargeable, irreversible and disposable).

The information shown below is for illustration only and does not form part of the AQA A-level specification.

Cell	Cell reactions	Cell e.m.f.	Electrodes	Uses
Leclanché or Dry Cell	**Reduction:** $MnO_2(s) + H_2O(l) + e^- \rightarrow MnO(OH)(s) + OH^-(aq)$ **Oxidation:** $Zn(s) \rightarrow Zn^{2+}(aq) + 2e^-$ **Cell reaction:** $2MnO_2(s) + 2H_2O(l) + Zn(s) \rightarrow Zn^{2+}(aq)$ $+ 2MnO(OH)(s) + 2OH^-(aq)$	1.5 V	**Reduction**: Graphite **Oxidation**: Zinc	Standard **'disposable'** cell
Alkaline cell	**Reduction:** $MnO_2(s) + H_2O(l) + e^- \rightarrow MnO(OH)(s) + OH^-(aq)$ **Oxidation:** $Zn(s) + 2OH^-(aq) \rightarrow Zn(OH)_2(s) + 2e^-$ **Cell reaction:** $2MnO_2(s) + 2H_2O(l) + Zn(s) \rightarrow Zn(OH)_2(s)$ $+ 2MnO(OH)(s)$	1.54 V	**Reduction**: MnO_2 **Oxidation**: Zinc	**'long-life'** or **'high power'** cell *'Longer life'*: zinc casing not subject to acid attack. *'High power'*: provides a large steady current.
Mercury cell	**Reduction:** $HgO(s) + H_2O(l) + 2e^- \rightarrow Hg(l) + 2OH^-(aq)$ **Oxidation:** $Zn(s) + 2OH^-(aq) \rightarrow Zn(OH)_2(s) + 2e^-$ **Cell reaction:** $Zn(s) + HgO(s) + H_2O(l) \rightarrow Zn(OH)_2(s) + Hg(l)$ Mercury is very toxic, so careful disposal is needed at an official site rather than simply 'dumping'.	1.36 V	**Reduction**: HgO **Oxidation**: Zinc	Appliances with very low power needs. Small size, low current drain, long life, as in hearing-aids, digital watches and thermometers.
Lithium cell	**Reduction:** $Li^+ + MnO_2(s) + e^- \rightarrow LiMnO_2(s)$ **Oxidation:** $Li(s) \rightarrow Li^+ + e^-$ **Cell reaction:** $Li(s) + MnO_2(s) \rightarrow LiMnO_2(s)$ Lithium has the most negative of all reduction potentials, so batteries with large e.m.f. values can be produced.	1.5 to 3 V	**Reduction**: MnO_2 **Oxidation**: Lithium	Digital cameras, calculators and watches. Very long life. Lithium iodide cells in pacemakers can last for up to 15 years.

Example

In the presence of ammonium chloride (commonly added to dry cells), the overall cell reaction is:

$$Zn(s) + 2MnO_2(s) + 4NH_4^+(aq) + 2OH^-(aq) \rightarrow$$
$$[Zn(NH_3)_4]^{2+}(aq) + 2MnO(OH)(s) + 2H_2O(l)$$

The $[Zn(NH_3)_4]^{2+}/Zn$ electrode has a standard potential of -1.03 V and the *cell e.m.f.* is 1.5 V.

(a) Write the half-equation for the $[Zn(NH_3)_4]^{2+}/Zn$ electrode.

(b) Deduce the half-equation for the reaction at the other electrode (the right-hand electrode) of this dry cell.

(c) On the assumption that it operates under standard conditions, determine the standard electrode potential of the other electrode of this dry cell.

Answer

(a) This is a two-electron reduction with $[Zn(NH_3)_4]^{2+}(aq)$ as the oxidised species and $Zn(s)$ as the reduced one; so the reduction half-equation is:
$$[Zn(NH_3)_4]^{2+}(aq) + 2e^- \rightarrow Zn(s) + 4NH_3(aq)$$

(b) Adding this to the overall cell equation (given above), and cancelling terms, gives:
$$2MnO_2(s) + 4NH_4^+(aq) + 2OH^-(aq) + 2e^- \rightarrow$$
$$4NH_3(aq) + 2MnO(OH)(s) + 2H_2O(l)$$
and, allowing for the Brønsted–Lowry acid–base reaction of NH_4^+ and OH^- forming ammonia and water, further cancellation gives:
$$2MnO_2(s) + 2NH_4^+(aq) + 2e^- \rightarrow 2NH_3(aq) + 2MnO(OH)(s)$$

So, in its simplest form, the half-equation is:
$$MnO_2(s) + NH_4^+(aq) + e^- \rightarrow NH_3(aq) + MnO(OH)(s)$$

(c) Using e.m.f. $= E_R - E_L$, with the $[Zn(NH_3)_4]^{2+}/Zn$ couple as the electrode on the left:
$$+1.5 = E_R - (-1.03), \text{ making } E_R = 1.5 - 1.03 = +0.47 \text{ V}$$

Comment

The ammonium chloride plays a pivotal role in ensuring that local concentrations of the battery components, those close to the electrodes, do not veer towards equilibrium conditions and thus decrease the e.m.f. of the cell.

Examiners' Notes

A galvanic cell is one with a positive e.m.f. (> 0) in which the spontaneous forward cell reaction can be used to provide electric current to an external circuit.

An electrolytic cell is one with a negative e.m.f., which makes the forward cell reaction impossible, but which is then made possible by the flow of an electric current from an external source.

Secondary cells

Secondary cells are *reversible cells*. They combine the two opposing functions of all reversible electrical cells, which are to provide current to an external circuit while discharging (*galvanic action*) and to use current from an external circuit while charging (*electrolytic action*). To be truly reversible, it is important that products resulting from both galvanic action and electrolytic action are not dispersed in the cell electrolyte but (being insoluble) remain attached to the cell electrodes.

Table 22
Secondary cells (rechargeable).
The information shown below is for illustration only and does not form part of the AQA A-level specification.

Cell	Cell reactions	Cell e.m.f.	Electrodes	Uses
Lead–acid	**Reduction on discharge:** $PbO_2(s) + 3H^+(aq) + HSO_4^-(aq) + 2e^-$ $\xrightleftharpoons[\text{charge}]{\text{discharge}} PbSO_4(s) + 2H_2O(l)$ $E^{\ominus} \approx +1.68$ V **Oxidation on discharge:** $PbSO_4(s) + H^+(aq) + 2e^-$ $\xrightleftharpoons[\text{discharge}]{\text{charge}} Pb(s) + HSO_4^-(aq)$ $E^{\ominus} \approx -0.36$ V **Cell reaction:** $PbO_2(s) + Pb(s) + 2H^+(aq) + 2HSO_4^-(aq)$ $\xrightleftharpoons[\text{charge}]{\text{discharge}} 2PbSO_4(s) + 2H_2O(l)$ $e.m.f. \approx +2.04$ V In motor cars, these batteries are discharged at very high current when used to power a starter motor, and are then 'trickle' charged by an alternator when there is no current demand on the battery.	2.04 V	**Reduction:** PbO_2 **Oxidation:** Lead	The workhorse of secondary cells, in daily use in battery form in motorcars. So called 'sealed' car batteries do not need any maintenance. They are 'valve regulated', as 'sealed' could present very serious hazards.
Nickel–cadmium	**Reduction on discharge:** $NiO(OH)(s) + 2H_2O(l) + 2e^-$ $\xrightleftharpoons[\text{charge}]{\text{discharge}} Ni(OH)_2(s) + 2OH^-(aq)$ $E^{\ominus} \approx +0.52$ V **Oxidation on discharge:** $Cd(OH)_2(s) + 2e^- \xrightleftharpoons[\text{discharge}]{\text{charge}} Cd(s) + 2OH^-(aq)$ $E^{\ominus} \approx -0.88$ V **Cell reaction:** $NiO(OH)(s) + Cd(s) + 2H_2O(l)$ $\xrightleftharpoons[\text{charge}]{\text{discharge}} 2Ni(OH)_2(s) + Cd(OH)_2(s)$ $e.m.f. \approx +1.4$ V	1.4 V	**Reduction:** $Ni(OH)_2$ **Oxidation:** Cadmium	All portable equipment needing high power, e.g. drills and chain saws.
Lithium ion	**Reduction on discharge:** $Li^+ + CoO_2 + e^- \xrightleftharpoons[\text{charge}]{\text{discharge}} LiCoO_2 \quad E^{\ominus} \approx +0.6$ V **Oxidation on discharge:** $Li^+ + e^- \xrightleftharpoons[\text{discharge}]{\text{charge}} Li \quad\quad E^{\ominus} \approx -3.0$ V **Cell reaction:** $Li + CoO_2 \xrightleftharpoons[\text{charge}]{\text{discharge}} LiCoO_2 \quad e.m.f. \approx +3.6$ V Lithium is very light and also has the highest negative standard electrode potential, so the e.m.f. is very high.	3.6 V	**Reduction:** Lithium cobaltate **Oxidation:** Graphite	Fastest growing rechargeable battery system, in common use in mobile phones, i-pods and lap-tops.

Characteristic features of some common secondary cells are shown in Table 22. The oxidation/reduction labels in the cell reactions refer to the discharge reactions. When charging, these reactions are reversed, replenishing the battery.

Fuel cells

Fuel cells use a supply of hydrogen or organic fuel, together with a supply of oxygen, to provide a source of electrical power. The simplest fuel cell uses energy from the reaction of hydrogen with oxygen to provide electrical power.

Fuel cells can replace the rather inefficient conversion (40%) into electricity from heat derived from combustion of hydrogen, with an electrochemical process that involves the continuous replenishment of fuel and oxygen at suitable electrodes to provide electrical power by a more direct method. Working at best attainable efficiencies, fuel cells can convert up to 85% or more of the chemical energy of combustion directly into electrical energy.

The **hydrogen–oxygen fuel cell** has two inert, porous, electrodes of nickel and nickel oxide (these permit the passage of the reactant gases as well as of the product, steam) immersed in a solution of hot aqueous potassium hydroxide, with gaseous hydrogen and oxygen bubbled into the cell continuously from opposite sides. The porous electrodes are impregnated with catalyst materials to speed up the electrode reactions and thus provide higher currents.

A schematic diagram is shown in Fig 7.

Fig 7
The hydrogen–oxygen fuel cell. Connections to an external circuit are shown as, in their absence, the hydrogen fuel cannot react with the oxygen. Electrons flowing in the external circuit are transferred from hydrogen to oxygen down the wire, rather than as a result of molecular collisions

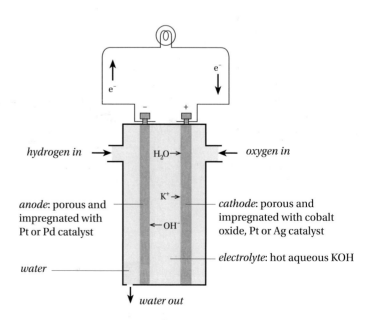

hydrogen in →

H_2O →

← oxygen in

K^+ →

← OH^-

anode: porous and impregnated with Pt or Pd catalyst

cathode: porous and impregnated with cobalt oxide, Pt or Ag catalyst

electrolyte: hot aqueous KOH

water

water out

The electrode reactions are:

Reduction: $O_2(g) + 2H_2O(l) + 4e^- \rightarrow 4OH^-(aq)$ $E^{\ominus}/V = +0.40$

Oxidation: $2H_2(g) + 4OH^-(aq) \rightarrow 4H_2O(l) + 4e^-$ $E^{\ominus}/V = +0.83$

Overall: $2H_2(g) + O_2(g) \rightarrow 2H_2O(l)$ e.m.f./V = +1.23

The overall reaction is entirely equivalent to direct combustion of hydrogen in oxygen.

The favourable enthalpy of combustion of hydrogen and oxygen is converted directly into electrical energy. Fuel cells, therefore, provide an attractive means of energy production, in particular because they do not produce effluents that increase global warming.

Although this output e.m.f. is based on standard conditions (298 K, 100 kPa), the hydrogen–oxygen fuel cell has only limited use under such conditions. The problem lies with the ability to produce appreciable electric current, and hence useable power. The diffusion rate of oxygen into the porous electrodes is not high at 298 K and, as a consequence, the current produced by the cell (which depends on the rate of production of electrons that, in turn, depends on the rate of conversion of gaseous oxygen to aqueous hydroxide ions) is low. To counteract this, it is usual to operate such cells at around 200 °C, with the higher temperature increasing the overall reaction rate (*kinetics*) and hence the available current.

Unfortunately, because the combustion of hydrogen is so very strongly exothermic, this increased temperature acts against the cell (*Le Chatelier*) and the e.m.f. falls. An increased pressure of 20–40 bar can compensate for the increased temperature (*Le Chatelier*), and at 200 °C and 40 bar, an e.m.f. of around 1.2 V can be achieved.

Hydrogen/oxygen fuel cells that operate at even higher temperatures (>500 °C) can achieve efficiencies in excess of 90%, but two notes of caution need to be expressed:

- Operating at a high temperature is sensible only if the heat generated can be used advantageously and not just dissipated. Lost heat seriously degrades the efficiency of the conversion process. Current thinking is moving towards *Combined Heat and Power (CHP)* systems, where excess heat can be used up, domestically, while power is being generated. Such systems are made even more cost-effective by feeding excess power, during 'idle' periods, into the electricity grid, thus running the electricity meter 'backwards' and decreasing bills.

Examiners' Notes

For clarity, it is useful to label the electrode half-reactions by their redox functions and then use the *addition method.*

Examiners' Notes

The chemistry of the hydrogen–oxygen fuel cell illustrates how chemists go about solving the unremitting problems of slow rates and low yields. In this case, the 'yield' is electric current, which itself involves the rate of flow of electrons. So, to get useful power from a fuel cell, it is necessary to have fast reactions (increased temperature and use of catalysts) that provide a good yield of product (from an increased pressure of reactants). The chemical product is water, itself not of any commercial value, but the more water produced in a shorter time, the more electrical current that can flow, and hence the greater the electrical power output of the fuel cell.

- While oxygen (in the form of air) is in plentiful and cheap supply, hydrogen is not. Hydrogen needs to be manufactured and one of the common sources is the *steam reforming* reaction on natural gas:

$$CH_4(g) + H_2O(g) \rightarrow CO(g) + 3H_2(g)$$

with extra hydrogen available from the lower temperature reaction of carbon monoxide and steam:

$$CO(g) + H_2O(g) \rightarrow CO_2(g) + H_2(g)$$

Overall : $CH_4(g) + 2H_2O(g) \rightarrow CO_2(g) + 4H_2(g)$

Both of these reactions demand an input of energy, so hydrogen is not produced without an overall energy deficit. Moreover, this method leaves behind it a large carbon footprint, as one mole or more (the process is not 100% efficient) of carbon dioxide results for every four moles of hydrogen produced.

Overall, the 90% efficiency of conversion can outweigh the energy demands of hydrogen production (though not so much the carbon footprint), so hydrogen/oxygen fuel cells are firmly in place as a future source of supply of electrical energy. The power output of batteries of such high-temperature fuel cells can now exceed 10 megawatts, an amazing source of clean power, with only water as the ultimate 'pollutant'.

However, a strong note of caution is still needed. Fuel cells can provide a method of electricity generation that has a zero-carbon footprint only if the hydrogen needed is manufactured by some means that does not itself produce carbon dioxide. Such means of production are available, most notably by the electrolysis of acidified water, but these processes are only truly 'green' if the electricity needed for electrolysis is itself generated by carbon-neutral sources such as wind, tide, hydroelectric, or nuclear power. Before fuel cells can be viewed as 'clean', the carbon footprint of the production of source materials must be considered alongside the fact that the cells themselves produce only water as a product and do not, at first sight, appear to pollute.

3.5.4 Transition metals

General properties of transition metals

The transition elements occupy the large central block of the Periodic Table, the **d block** (see *Collins Student Support Materials: Unit 1 – Foundation Chemistry*, section 3.1.4). However, in the present definition of a transition element, not all d-block elements are considered to be transition elements.

> **Definition**
>
> A *transition element* is an element having an incomplete d (or f) sub-level either in the element or in one of its common ions.

The 3d sub-level can contain up to ten electrons; there are therefore ten elements in the first row of the d block from scandium to zinc. Titanium is the second element, the electron arrangement of which can be deduced from the Periodic Table as:

Ti $\quad$ [Ar]$3d^24s^2$

This electron arrangement has an incomplete d sub-level, so that titanium is a transition element. Across the Periodic Table, the d sub-level is progressively filled until copper, which has the electron arrangement:

Cu $\quad$ [Ar]$3d^{10}4s^1$

This electron arrangement does not have an incomplete d sub-level. However, one of the common ions of copper is Cu^{2+}, which does have an electron arrangement with an incomplete d sub-level:

Cu^{2+} $\quad$ [Ar]$3d^9$

Copper is therefore a transition element.

After copper, the next element is zinc, which has the electron arrangement:

Zn $\quad$ [Ar]$3d^{10}4s^2$

and it forms only one common ion, which has the electron arrangement:

Zn^{2+} $\quad$ [Ar]$3d^{10}$

Neither in the element nor in its common ion does zinc have an incomplete d sub-level so that zinc is not classed as a transition element.

In working out the electron arrangement of a transition-metal ion, the *outer s electrons are always lost first* from the electron arrangement of the metal. Transition-metal ions and transition metals in compounds do not have any outer s electrons; it is the partly filled d sub-level which is responsible for the characteristic properties of the transition-metal ions in their compounds.

Examiners' Notes

Scandium [Ar]$3d^14s^2$ is the first transition element, but is not included in the A-level specification.

Examiners' Notes

Note that [Ar]$4s^23d^2$ is equally acceptable as a representation.

Examiners' Notes

This electron arrangement is more stable than [Ar]$3d^94s^2$.

Characteristic properties of transition elements are:

- formation of complexes
- formation of coloured ions
- variable oxidation states
- catalytic activity.

Complex formation

Complex compounds contain a central atom surrounded by ions or molecules, both of which are called ligands.

Examiners' Notes

Strictly, a ligand is a lone-pair donor only when it is actually bonded to a metal ion.

> **Definition**
>
> A **ligand** is any atom, ion or molecule which can donate a pair of electrons to a metal ion.

The ability to donate a pair of electrons means that a **ligand**, a **Lewis base** and a **nucleophile** are equivalent.

Examiners' Notes

Lewis bases are discussed more fully in section 3.5.5.

When a complex compound is formed, ligands donate an electron pair to a metal ion to form a **co-ordinate bond** with the metal.

> **Definition**
>
> The number of atoms bonded to a metal ion is called the **co-ordination number**.

Examiners' Notes

In a complex compound, the co-ordination number of the metal differs from its oxidation state.

In the hexaaquacopper(II) ion, each water molecule donates an electron pair (one of its lone pairs) to the copper(II) ion to form an octahedral complex:

$$CuSO_4(s) \xrightarrow{\;H_2O\;} [Cu(H_2O)_6]^{2+}$$

copper(II) sulfate   hexaaquacopper(II) ion

white   blue

Co-ordination number of Cu = 6

Oxidation state of Cu = +2

For ligands with only one donor atom, the co-ordination number is the number of ligands bonded to the metal ion; this is not true for multidentate ligands (see page 55).

It is important to remember that, while the hexaaqua complex is an ion, the bonds within the complex, i.e. the Cu—O and the O—H bonds, are covalent (see section 3.5.5 for a diagram of the hexaaqua ion). Reactions of the complex ion involve the breaking of one or both of these types of bond.

A ligand such as water, which has only one atom that can donate a pair of electrons, and which consequently bonds through one atom only, is said to be **unidentate**. Unidentate ligands include:

H_2O, NH_3, Cl^-, OH^- and CN^-

Although several of these species have more than one lone pair of electrons, each ligand donates only one lone pair on co-ordination.

Ligands which contain two donor atoms, and which consequently are able to bond to a metal ion through two atoms, are called **bidentate**. Bidentate ligands include:

ethane-1,2-diamine (*ethylenediamine* or *en*), $H_2NCH_2CH_2NH_2$ (Fig 8), which bonds through *two* nitrogen atoms, and the ethanedioate (*oxalate*) ion, $C_2O_4{}^{2-}$ (Fig 9), which bonds through *two* oxygen atoms.

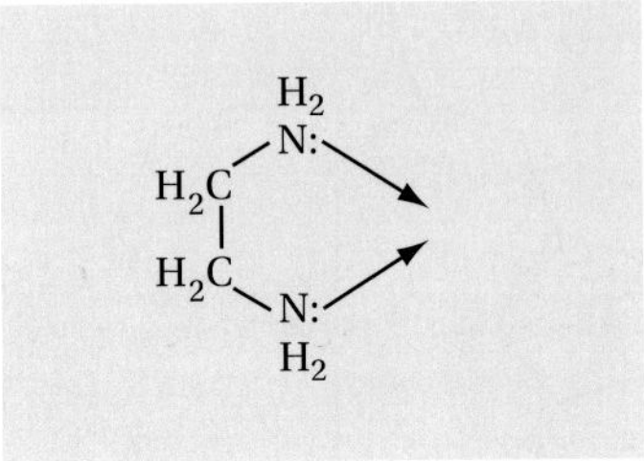

Fig 8
Ethane-1,2-diamine
(*ethylenediamine*)

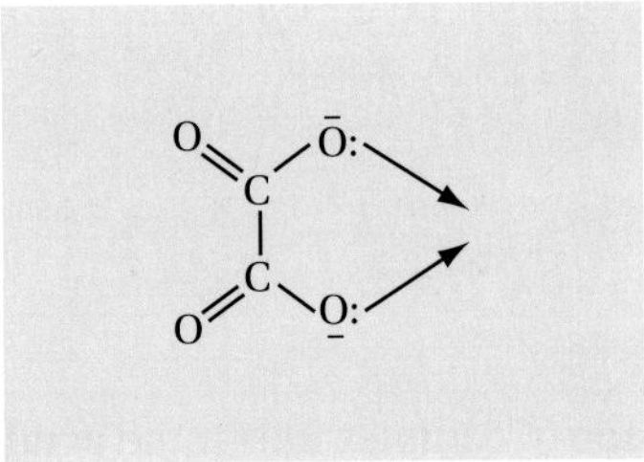

Fig 9
The ethanedioate (*oxalate*) ion

Some ligands contain many donor atoms and are said to be **multidentate**. Typical of these is the anion derived from bis[di(carboxymethyl)amino]ethane, commonly known as ethylenediaminetetraacetic acid or H_4EDTA. The anion **EDTA^{4-}** is able to bond to metal ions from six donor atoms.

The structure of the EDTA^{4-} anion is shown in Fig 10.

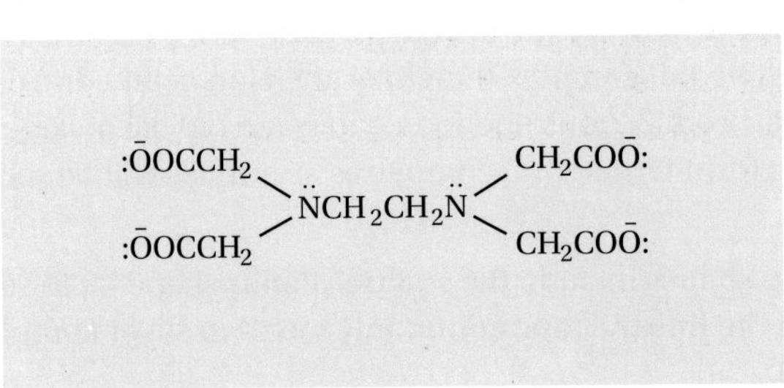

Fig 10
The structure of the EDTA^{4-} anion

Examiners' Notes

EDTA can be used to estimate many metals volumetrically. An indicator is used which forms a weak complex with the metal ion. When all the free metal ion has been complexed with EDTA, the indicator is displaced from its weak metal complex and a new colour is seen at the end-point.

EDTA uses the lone pairs on its six donor sites (4O and 2N) and forms 1:1 complexes with metal(II) ions, for example:

$$[Cu(H_2O)_6]^{2+} + EDTA^{4-} \rightarrow [Cu(EDTA)]^{2-} + 6H_2O$$

Blood contains a red iron(II) complex called haem. In this complex, shown in Fig 11, the iron is bonded to four nitrogen atoms in a plane within a large organic molecule called porphyrin (which is therefore a quadridentate ligand). Octahedral co-ordination of the iron involves a fifth nitrogen atom, above this plane, from a protein called globin, with the sixth position, below the plane, occupied by either molecular oxygen or the oxygen atom of a water molecule.

Fig 11
An iron(II) ion at the centre of a porphyrin ring in haem. Various organic side chains are attached to the five-membered rings

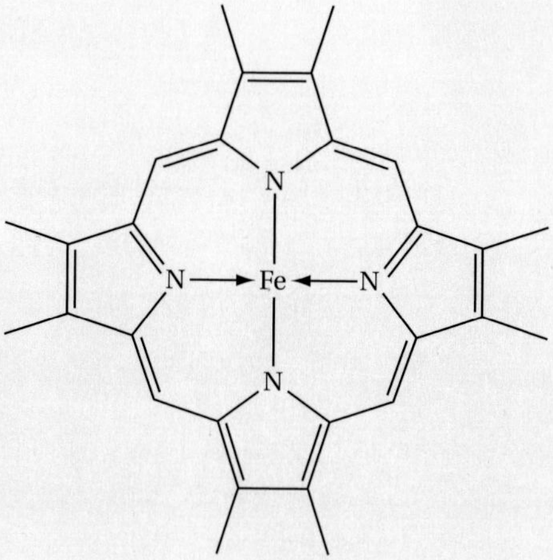

Shapes of complex ions

The most common shape of complex ions is the **octahedral** shape formed, for example, when water molecules co-ordinate to a copper(II) ion. All the transition metals of the first-row transition series form octahedral hexaaqua ions with water. Ammonia is another ligand with no charge; it is similar in size to water and readily forms octahedral complexes with these metals, as will be seen later in section 3.5.5.

The next most common shape encountered in complexes is **tetrahedral**. If the ligand is large and negatively charged (so that there is inter-ligand repulsion as well as bond-pair repulsion in the complex), then it may not be possible to fit six ligands around the central metal ion in a stable complex, so the tetrahedral arrangement is preferred. Ligands are further apart in a tetrahedral complex than they are in an octahedral complex. Chloride ions, bromide ions and iodide ions are typical of large anions which form tetrahedral complexes; examples will be seen later in section 3.5.5.

Examiners' Notes

Fluoride ions do form octahedral complexes, e.g. $[AlF_6]^{3-}$ and $[FeF_6]^{3-}$, because the F^- ion is smaller than the Cl^- ion.

Less common shapes include the **square-planar** shape in, for example, cisplatin and the **linear** shape commonly found in silver complexes (see page 68).

The arrangements of the bonds in the different shapes of complexes are shown in Fig 12.

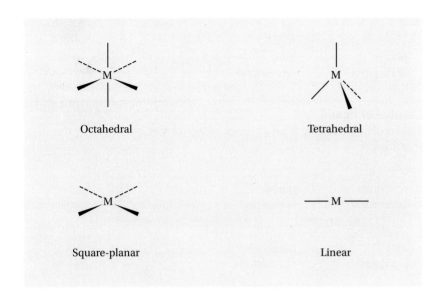

Octahedral

Tetrahedral

Square-planar

Linear

Fig 12
The shapes of complex ions

Formation of coloured ions

Many of the coloured solutions which are seen around the laboratory owe their colour to the presence of a transition metal. Indeed, along with other evidence, the colour of a solution can be used to identify the presence of a particular transition metal.

When a colour change occurs in the reaction of a transition-metal ion, there is a change in at least one of the following factors:

- oxidation state
- co-ordination number
- ligand.

Often there is a change in more than one of these factors.

Change of oxidation state

Examples of a change in oxidation state being responsible for the colour change are:

$[Fe^{II}(H_2O)_6]^{2+}$ → $[Fe^{III}(H_2O)_6]^{3+}$
green   very pale violet

$[Cr^{III}(H_2O)_6]^{3+}$ → $[Cr^{II}(H_2O)_6]^{2+}$
red-violet   blue

Change of co-ordination number

A change in co-ordination number (i.e. the number of nearest neighbours) is most usually achieved by changing ligands. Examples of

> **Examiners' Notes**
>
> See the hydrolysis reactions later for an explanation of the brown colour seen in iron(III) solutions. Similarly the green colour (emerald) usually seen in chromium(III) compounds is due to hydrolysis or substitution by other anions, as in $[Cr(H_2O)_4Cl_2]^+$.

colour changes arising from this type of change are:

$[Cu(H_2O)_6]^{2+}$ → $[CuCl_4]^{2-}$
blue   yellow-green
octahedral   tetrahedral

$[Co(H_2O)_6]^{2+}$ → $[CoCl_4]^{2-}$
pink   blue
octahedral   tetrahedral

Change of ligand

Colour changes arising from a change of ligand only are shown by the reactions:

$[Cr(H_2O)_6]^{3+}$ → $[Cr(NH_3)_6]^{3+}$
red-violet   purple
octahedral   octahedral

$[Cu(H_2O)_6]^{2+}$ → $[Cu(NH_3)_4(H_2O)_2]^{2+}$
blue   blue-violet
octahedral   octahedral

Origin of colour

Colour arises when a molecule or an ion absorbs visible light and enters a higher or *excited* energy state. If only a part of the visible spectrum is absorbed, then the eye detects those frequencies which remain and a colour is seen.

Energy is absorbed when an electron is promoted from the ground state to a higher energy level, as shown in Fig 13. The energy difference between the two levels involved is given by:

$$E_2 - E_1 = \Delta E = h\nu$$

where h is **Planck's constant** and ν is the frequency of the absorbed radiation.

Examiners' Notes

In some reactions, the colour change arises as a result of all three changes, e.g.

$[CrO_4]^{2-}$ → $[Cr(H_2O)_4Cl_2]^+$
Cr(VI)   Cr(III)
yellow   green
tetrahedral   octahedral

and

$[Mn(H_2O)_6]^{2+}$ → $[MnO_4]^-$
Mn(II)   Mn(VII)
very pale pink   purple
octahedral   tetrahedral

Essential Notes

$h = 6.63 \times 10^{-34}$ J s

Fig 13
Absorption of light giving rise to colour. The symbol ↑ represents an electron

Essential Notes

If both levels involved are levels of d orbitals, the change is called a *d–d transition* and the associated colour is not very intense. If, however, the transition involves an electron moving from a ligand orbital to a metal (d) orbital (or vice-versa), the transition is called *charge-transfer* and an intense colour results.

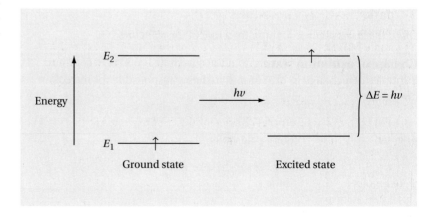

Visible spectrophotometry

The intensity of colours shown by a solution of a transition-metal ion can be used to determine the concentration of that ion; this is done using a spectrophotometer.

In a **visible spectrophotometer**, light of increasing (or decreasing) frequency is passed through the sample; the emergent light is received in a detector and recorded. The *amount of light absorbed is proportional to the concentration of the absorbing species* in the solution under test. By measuring how much light is absorbed by a solution at a particular frequency, the concentration of the absorbing species can be determined.

So sensitive are these colours to a change in the concentration of the ion, that a very simple spectrophotometer can be used to measure the intensity of absorption at the particular frequency absorbed. Such spectrophotometers are called **colorimeters** (because they measure the intensity of colours).

Variable oxidation states

A characteristic property of the transition elements is their ability to change oxidation state in chemical reactions.

In an oxidation–reduction or 'redox' reaction, the transition-metal ion is oxidised or reduced, i.e. it changes its oxidation state. Because transition-metal ions show variable oxidation states, many redox reactions occur.

Metallic zinc is a good reducing agent and reduces aqueous transition-metal ions to low oxidation states (see section 3.5.3). Zinc reacts with mineral acids to form zinc(II) ions and releases two electrons for the reduction:

$$Zn \rightarrow Zn^{2+} + 2e^-$$

Reductions with zinc proceed through the oxidation states of the metal in sequence, so that each different colour of the metal ion in every oxidation state can often be seen. A good example of this behaviour is seen in the reduction of chromium(VI).

Oxidation states of chromium

Chromium shows its highest oxidation state of +6 in the yellow chromate(VI) ion, CrO_4^{2-}, and the orange dichromate(VI) ion, $Cr_2O_7^{2-}$. The dichromate ion is formed from the chromate ion in acid solution:

$$2CrO_4^{2-} + 2H^+ \rightarrow Cr_2O_7^{2-} + H_2O$$
$$\text{yellow} \qquad\qquad \text{orange}$$

The reduction of dichromate(VI) ions by zinc and hydrochloric acid proceeds through the following stages:

Cr(VI)	$Cr_2O_7^{2-}$		orange
↓	↓		↓
Cr(III)	$Cr^{3+}(aq)$	i.e. $[CrCl_2(H_2O)_4]^+$	green
↓	↓		↓
Cr(II)	$Cr^{2+}(aq)$	i.e. $[Cr(H_2O)_6]^{2+}$	blue

Essential Notes

The **Beer–Lambert law:** $A = \varepsilon cl$ links absorbance (A) to concentration (c), cell path-length (l) and molar absorption coefficient (ε).

Examiners' Notes

The colorimeter is a relatively cheap instrument that is used in industry and research to measure the concentration of coloured species in a rapid and non-destructive manner. Colorimeters can be operated remotely and the output fed to a computer for rapid analysis or data storage.

Examiners' Notes

The ending -*ate* means that the metal is present in a negatively-charged ion (an anion).

Examiners' Notes

The chromate/dichromate change is not a redox reaction; it is reversed in alkaline solution:

$$Cr_2O_7^{2-} + 2OH^- \rightarrow 2CrO_4^{2-} + H_2O$$

It is an acid–base *equilibrium* reaction.

Examiners' Notes

The red-violet colour of hexaaquachromium(III) is seen in solid compounds containing this ion, e.g. 'chrome alum'. This colour is sometimes described as *ruby*; natural gemstones of ruby owe their colour to the octahedral arrangement of oxygen atoms around chromium(III). The gemstone *emerald* also owes its colour to six-coordinate chromium(III) ions, but the six ligands are no longer identical.

In order to reach the chromium(II) state, it is necessary to keep air away from the reduction. When air is admitted to the blue chromium(II) solution, the colour rapidly changes to green due to the presence of chromium(III). Once again it is clear that, in hydrochloric acid, some substitution of the chromium(III)-aqua ion has taken place. Even in sulfuric acid, substitution of sulfate ions into the hexaaqua ion occurs, so that the red-violet (ruby) colour of the $[Cr(H_2O)_6]^{3+}$ ion is not seen.

Oxidation states of manganese

Manganese, the next element along the first transition series, shows its maximum oxidation state of +7 in the manganate(VII), MnO_4^-, ion. This ion is a powerful oxidising agent which is reduced directly to the manganese +2 state by zinc, by iron, or even by iron(II) ions in acid solution.

The $[Mn(H_2O)_6]^{2+}$ ion is a very pale pink colour and appears colourless in aqueous solution so that, when manganate(VII) ions are reduced to Mn(II) ions, the colour change seen is from purple to colourless:

Mn(VII)	MnO_4^-	purple
↓	↓	↓
Mn(II)	$[Mn(H_2O)_6]^{2+}$	colourless

Redox titrations

Redox reactions can be used in quantitative volumetric analysis. The end-point, when the reaction is complete, can be determined in a variety of ways, including the use of indicators or electrical methods. The use in **redox titrations** of two common quantitative oxidising agents is discussed below.

Potassium manganate(VII) titrations

When potassium manganate(VII) is used as an oxidising agent in acidic solution, no additional indicator is required. Because the reduction product, $[Mn(H_2O)_6]^{2+}$, is essentially colourless, no colour is seen until an excess of potassium manganate(VII) has been added, at which point the presence of even a slight excess of the dark purple manganate(VII) ion is readily seen and is used to indicate complete reaction. Knowing the initial and final species present, MnO_4^- and Mn^{2+}, the following half-equation can be constructed:

$$MnO_4^- + 8H^+ + 5e^- \rightarrow Mn^{2+} + 4H_2O$$

The presence of an excess of acid for this reaction is essential and the choice of the actual acid used is important. The acid must:

- be strong, because a high concentration of hydrogen ions is needed

- not be an oxidising agent, because an oxidising agent might react with the reducing agent being estimated

- not be a reducing agent, because reducing agents will be oxidised by the manganate(VII) ions.

Examiners' Notes

If insufficient acid is present, a brown colouration of insoluble MnO_2 is seen.

A review of common laboratory acids enables the correct choice of acid to be made:

- it *cannot* be hydrochloric acid, as this can be oxidised to chlorine
- it *cannot* be nitric acid, as this is an oxidising agent
- it *cannot* be *concentrated* sulfuric acid, as this too is an oxidising agent
- it *cannot* be ethanoic acid, as this is a *weak* acid which produces too low a concentration of H^+ ions.

Dilute sulfuric acid is, therefore, the main choice among common laboratory acids.

The usual procedure is to add a standard solution of potassium manganate(VII) from a burette to a conical flask containing a solution of the reducing agent which has been acidified by the addition of an excess of dilute sulfuric acid. As mentioned above, no additional indicator is required for this titration.

Consider, for example, the use of potassium manganate(VII) to estimate the concentration of iron(II) ions in solution. Knowing that the oxidation of iron(II) to iron(III) is a one-electron change and that the conversion of manganese(VII) into manganese(II) is a five-electron change, it is easy to deduce that the complete reduction of one mole of manganese(VII) to manganese(II) will require five moles of iron(II). The two half-equations and the overall equation for these processes are:

$$MnO_4^- + 8H^+ + 5e^- \rightarrow Mn^{2+} + 4H_2O$$
$$5Fe^{2+} \rightarrow 5Fe^{3+} + 5e^-$$
$$\overline{5Fe^{2+} + MnO_4^- + 8H^+ \rightarrow 5Fe^{3+} + Mn^{2+} + 4H_2O}$$

Potassium dichromate(VI) titrations

For redox titrations using potassium dichromate(VI) as the oxidant, an indicator is required. This is because the colour change is from orange $Cr_2O_7^{2-}$ to green Cr^{3+} and a small excess of orange is hard to detect in the green solution. One of the more commonly-used indicators is sodium diphenylaminesulfonate, which turns from colourless to purple at the end-point.

Potassium dichromate(VI) can also be used to estimate iron(II) ions in solution. Because chromium is reduced from the +6 to the +3 oxidation state, the reacting ratio is one mole of chromium(VI) to three moles of iron(II):

$$Cr_2O_7^{2-} + 14H^+ + 6e^- \rightarrow 2Cr^{3+} + 7H_2O$$
$$6Fe^{3+} \rightarrow 6Fe^{3+} + 6e^-$$
$$\overline{6Fe^{2+} + Cr_2O_7^{2-} + 14H^+ \rightarrow 6Fe^{3+} + 2Cr^{3+} + 7H_2O}$$

Again, such titrations are normally carried out with the dichromate(VI) solution in a burette and the iron(II) solution in a conical flask.

As with manganate(VII) titrations, H^+ is a reactant, so the acid must be present in excess. For titrations involving dichromate(VI), the H^+ ions can be provided either by dilute sulfuric acid or by dilute hydrochloric acid.

Examiners' Notes

Note that there are two moles of chromium(VI) in one mole of dichromate(VI).

It is possible to use hydrochloric acid in this case because, unlike the MnO_4^-/Mn^{2+} couple, the potential of the $Cr_2O_7^{2-}/Cr^{3+}$ couple is not positive enough to oxidise chloride ions to chlorine, so that the concentration of iron(II) ions on their own can be measured without interference from chloride ions.

Below are two examples which involve the estimation of iron(II) using each of these volumetric oxidising agents. The first example below shows the analysis by titration with potassium manganate(VII) solution of a sample of lawn sand which contains iron(II) sulfate.

Examiners' Notes

Typical laboratory experiments include:

- determining the percentage of iron(II) in commercial lawn sand or iron tablets
- determining the percentage of sulfite ions in a given solid, or in wine to which it has been added as a preservative.

Example

3.00 g of a lawn sand containing an iron(II) salt were shaken with dilute sulfuric acid. The resulting solution required 25.00 cm^3 of a 0.0200 mol dm^{-3} solution of $KMnO_4$ to oxidise all the Fe^{2+} ions in the solution to Fe^{3+} ions. Calculate the percentage by mass of Fe^{2+} ions in the lawn sand.

Method

Calculate the number of moles of manganate(VII) that have reacted, and then use the stoichiometric equation to relate this number to the number of moles of Fe(II) that have been oxidised. Use the relative atomic mass of iron to convert moles into mass and then express the mass as a percentage of the original sample.

Answer

$$\text{moles } KMnO_4 = \frac{0.0200 \times 25.00}{1000} = 5.00 \times 10^{-4} \text{ mol}$$

There are *five* moles of Fe^{2+} for every mole of MnO_4^- used, so

$$\text{moles } Fe^{2+} = 5 \times 5.00 \times 10^{-4} = 2.50 \times 10^{-3} \text{ mol}$$

Multiplying the number of moles by the mass of one mole of Fe^{2+} gives the mass of Fe^{2+} present in 3.00 g of lawn sand:

$$\text{mass of } Fe^{2+} \text{ present} = 55.8 \times 2.5 \times 10^{-3} = 0.140 \text{ g}$$

Hence, the percentage by mass of Fe^{2+} ions present in the sample is

$$\frac{0.140 \times 100}{3.00} = 4.65\%$$

Comment

Note that it is not necessary to filter off the sand in this analysis, because it is inert and does not interfere with the titration.

Note also that *rounding off* the mass of Fe^{2+} to 3 significant figures *before* calculating the percentage gives a slightly different, and less accurate, answer.

The following example shows the analysis by titration with potassium dichromate(VI) solution of a medicinal iron tablet which contains iron(II) sulfate hydrate and an inert filler.

Example

When a medicinal iron tablet weighing 0.940 g was dissolved in dilute sulfuric acid and the resulting solution was titrated with a 0.0160 mol dm^{-3} solution of $K_2Cr_2O_7$, exactly 32.5 cm^3 of the potassium dichromate(VI) solution were required to reach the end-point. Calculate the percentage by mass of Fe^{2+} in the tablet.

Method

Calculate the number of moles of dichromate(VI) that have reacted, and then use the stoichiometric equation to relate this number to the number of moles of Fe(II) that have been oxidised. Use the relative atomic mass of iron to convert moles into mass and then express the mass as a percentage of the original tablet.

Answer

$$\text{moles } Cr_2O_7{}^{2-} = \frac{32.5 \times 0.0160}{1000} = 5.20 \times 10^{-4}$$

There are *six* moles of Fe^{2+} for every mole of $Cr_2O_7{}^{2-}$ used, so

moles of $Fe^{2+} = 6 \times 5.20 \times 10^{-4} = 3.12 \times 10^{-3}$

Multiplying the number of moles by the mass of one mole of Fe^{2+} gives the mass of Fe^{2+} present in the 0.940 g tablet:

mass of Fe^{2+} is $55.8 \times 6 \times 5.20 \times 10^{-4} = 0.174$ g

Hence, the percentage by mass of Fe^{2+} ions in the tablet is:

$$\frac{0.174 \times 100}{0.940} = 18.5\%$$

Comment

Iron is needed by the body for good health. It is an essential component of haem, which is involved in the transport of oxygen by the blood. A lack of iron leads to the condition known as anaemia. Anaemia sufferers take iron tablets; these tablets are also given to some blood donors to restore their iron levels after giving blood.

Oxidation in alkaline solution

In the above titrations, the transition metal in a high oxidation state has been reduced in acid solution. In alkaline solution, transition-metal ions in low oxidation states are readily oxidised to higher oxidation states. For example, if a solution of Fe^{2+} ions is to be stored without being oxidised, it must be acidified. In alkaline solution, the precipitated green $Fe(OH)_2$ oxidises rapidly to form brown $Fe(OH)_3$.

Consider the species present:

$[M(H_2O)_6]^{2+}$	$[M(H_2O)_4(OH)_2]$	$[M(OH)_4]^{2-}$
acid solution	neutral	alkaline solution
harder to oxidise		easier to oxidise

Oxidation is electron loss. It is easier to remove an electron from a neutral or negatively-charged species than from a positively-charged one.

Examiners' Notes

Reduction occurs most readily in acid solution.

Oxidation occurs most readily in alkaline solution.

To prepare a metal complex which contains the metal in a high oxidation state, the general procedure is:

- first, add an alkali
- then add an oxidising agent.

This method is applied in the examples which follow.

The addition of an excess of sodium hydroxide to a solution of a chromium(III) salt gives a green solution containing the green chromate(III) ion (see section 3.5.5). When this solution is treated with hydrogen peroxide, a yellow solution containing chromate(VI) ions is formed. Thus, chromium(III) is readily oxidised to CrO_4^{2-} by hydrogen peroxide in alkaline solution:

$$2[Cr(OH)_6]^{3-} + 3H_2O_2 \rightarrow 2CrO_4^{2-} + 2OH^- + 8H_2O$$

Often, oxygen in the air can act as the oxidising agent; it is capable of oxidising $Fe(OH)_2$ to $Fe(OH)_3$, $Mn(OH)_2$ to $Mn(OH)_3$ and $Co(OH)_2$ to $Co(OH)_3$. Similar oxidations occur when hydrogen peroxide is added. The metal 2+ ions can be stabilised against atmospheric oxidation, however, by keeping them in acidic solution (see storage of iron(II) solutions on page 63).

The addition of aqueous ammonia to a solution of a cobalt(II) salt gives a green-blue precipitate of cobalt(II) hydroxide. If this suspension is shaken in air, oxidation to the brown cobalt(III) hydroxide takes place. If, however, the cobalt(II) hydroxide is treated with an excess of concentrated aqueous ammonia in the absence of air, a pale brown (straw-coloured) solution of the cobalt(II) hexaammine is formed. If air is then admitted, or hydrogen peroxide added, the hexaammine is rapidly oxidised to a dark brown mixture containing the hexaamminecobalt(III) ion:

$$[Co(H_2O)_6]^{2+} \xrightarrow{NH_3} [Co(OH)_2(H_2O_4)] \xrightarrow{\text{excess }NH_3} [Co(NH_3)_6]^{2+}$$

pink	green–blue	pale brown
octahedral	precipitate	octahedral

$$\downarrow O_2$$

$$[Co(NH_3)_6]^{3+}$$

yellow
octahedral

Catalysis

A **catalyst** (see *Collins Student Support Materials: Unit 2 – Chemistry in Action*, section 3.2.2) makes reactions go faster, i.e. it increases the rate of attainment of equilibrium. Catalysts are classified according to whether they act in a different phase from, or in the same phase as, the reactants.

Definition

*A catalyst which acts in a **different phase** from the reactants is called a **heterogeneous** catalyst.*

*A catalyst which acts in the **same phase** as the reactants is called a **homogeneous** catalyst.*

Because the success of an industrial process depends upon making the maximum amount of product in the shortest possible time, catalysis is at the heart of the chemical manufacturing industry. Most of the catalysts used in industry are transition metals or their compounds.

Heterogeneous catalysis

Heterogeneous catalysis depends on at least one reactant being adsorbed onto the catalyst surface (usually by weak chemical bonds), and being modified into a state which makes reaction more likely.

Sometimes **adsorption** onto the surface can weaken reactant bonds, or induce a reactant molecule to break up into very reactive fragments; but equally, adsorption may be responsible for holding a reactant molecule on the surface in exactly the right configuration to make reaction easier.

In simple gas reactions such as:

$A(g) + B(g) \rightarrow$ products

catalysis leading to an increase in reaction rate can come about if:

- **A** is adsorbed on the surface, becomes more accessible to collisions with **B**, and so reacts more readily than by random collisions in the gas phase. *Collision frequency increases.*

- **A** is held on the surface of the catalyst in a particularly favourable and highly reactive configuration. Unadsorbed **B** collides with reactive **A**. *Activation energy decreases.*

- **A** is adsorbed onto the surface and undergoes internal bond-breaking or rearrangement. Reactive fragments can easily react. *Activation energy decreases.*

Making catalysts more effective

Heterogeneous catalysts act through surface properties. Increasing the surface area available for reaction is always worthwhile. A wafer-thin metal foil is much more active catalytically than is the same mass of metal in a solid lump. Expensive catalysts are frequently spread very thinly or impregnated onto an **inert support medium** in order to maximise the **surface-to-mass ratio**. Solids which have large surface areas in their own right (e.g. silica) are useful as support media, especially if ground into very fine powders to maximise the surface area.

The practice of increasing surface area leads to more effective catalysis and also to the much more effective use of high-cost catalytic agents. *Loss of catalyst* from a support can greatly increase the inherent cost of a catalytic process, not least by decreasing the overall efficiency of the reaction.

Some catalysts can combine surface and support functions. The support itself may show catalytic activity and such **mixed catalysts** can often perform dual functions.

The most useful catalysts, such as Pt and Rh, are also among the most expensive. **Catalytic converters** in cars are designed to give a *high surface-area* of active catalyst and use rhodium spread thinly on a cheap ceramic support. Modern catalysts in car exhaust converters reduce noxious emissions to a tolerable level.

Manufacture of methanol

Chromium(III) oxide, Cr_2O_3, in a mixture with zinc and/or copper(II) oxides, is used as a catalyst in the synthesis of methanol from carbon monoxide and hydrogen:

$$CO + 2H_2 \rightarrow CH_3OH$$

The Haber and Contact processes

Two large-scale industrial processes use heterogeneous catalysis by a transition metal or one of its compounds. In the production of ammonia from nitrogen and hydrogen by the **Haber process**, iron is the catalyst. In the manufacture of sulfuric acid by the **Contact process**, the conversion of sulfur dioxide into sulfur trioxide is catalysed by vanadium(V) oxide. The sulfur trioxide reacts with water to give sulfuric acid. These two processes are summarised below.

Haber process : $\quad N_2(g) + 3H_2(g) \xrightleftharpoons{\;Fe(s)\;} 2NH_3(g)$

Contact process : $\quad 2SO_2(g) + O_2(g) \xrightleftharpoons{\;V_2O_5(s)\;} 2SO_3(g)$

The importance of the variable oxidation state of vanadium in the Contact process can be seen by considering how the catalyst works. Sulfur dioxide is oxidised to sulfur trioxide by vanadium(V) oxide:

$$SO_2 + V_2O_5 \rightarrow SO_3 + V_2O_4$$

The vanadium(V) has thus been reduced to vanadium(IV), which is then oxidised back to vanadium(V) oxide by oxygen:

$$2V_2O_4 + O_2 \rightarrow 2V_2O_5$$

Thus, the vanadium(V) oxide is unchanged at the end of the reaction and acts as a catalyst by use of the two oxidation states of vanadium, +5 and +4. The presence of the vanadium(V) oxide enables the reaction to proceed by a different route with a lower activation energy.

Catalyst poisons

Surface-active catalysts are especially prone to *poisoning*. For example, catalytic converters for car exhausts are readily poisoned by lead in *anti-knock* additives.

Poisoning occurs when unwanted contaminants, or waste products, become adsorbed too strongly onto the surface. Transition-metal catalysts, for example, are especially prone to poisoning by sulfur compounds, which form inactive surface sulfides. The effect is cumulative. The sulfur content of feedstock gases in metal-catalysed reactions must be reduced to a minimum.

Homogeneous catalysis

Reactions which are catalysed homogeneously by a transition-metal ion usually involve a change in oxidation state of the transition metal during catalysis. Such reactions proceed through the formation of an intermediate species, which can form because the metal has variable oxidation states.

The oxidation of iodide ions by peroxodisulfate(VI) ions catalysed by iron ions

The reaction:

$$S_2O_8^{2-} + 2I^- \rightarrow I_2 + 2SO_4^{2-}$$

is very slow even though it is energetically very favourable. Both ions are negatively charged and so are unlikely to make effective collisions with each other. The reaction occurs rapidly, however, when iron(II) ions are added. These cationic species can make effective collisions with anions. Iron(II) ions are oxidised by peroxodisulfate(VI) ions; the iron(II) becomes iron(III) which, in turn, can rapidly oxidise iodide ions to iodine:

$$2Fe^{2+} + S_2O_8^{2-} \rightarrow 2Fe^{3+} + 2SO_4^{2-}$$
$$\underline{2Fe^{3+} + 2I^- \rightarrow 2Fe^{2+} + I_2}$$
$$S_2O_8^{2-} + 2I^- \rightarrow I_2 + 2SO_4^{2-}$$

The variable oxidation state of iron enables the reaction to proceed by an alternative route with a lower activation energy.

Examiners' Notes

The two steps of the catalysed reaction can occur in either order, so that iron(III) ions can also be used as the catalyst.

Autocatalysis by manganese ions in the reaction between ethanedioate ions and manganate(VII) ions

When a solution of potassium manganate(VII), from a burette, is run into a hot, acidified solution containing ethanedioate ions, the purple colour is not decolourised immediately in the early stages of the titration. Once the initial purple colour produced by the addition of the first few drops of manganate(VII) solution has been discharged, however, further addition of the oxidant leads to immediate decolourisation. This rapid decolourisation continues until the end-point is reached.

The overall reaction is:

$$2MnO_4^- + 16H^+ + 5C_2O_4^{2-} \rightarrow 2Mn^{2+} + 10CO_2 + 8H_2O$$

The reaction between MnO_4^- ions and $C_2O_4^{2-}$ ions is slow; as in the previous example, both ions are negatively charged and so are less likely to make effective collisions. However, the reaction is catalysed by Mn^{2+} ions. Once some Mn^{2+} ions have been formed in the early stages of the titration, they can react with MnO_4^- ions to form a few Mn^{3+} ions:

$$4Mn^{2+} + MnO_4^- + 8H^+ \rightarrow 5Mn^{3+} + 4H_2O$$

These Mn^{3+} ions can then react with $C_2O_4^{2-}$ ions to liberate CO_2 and re-form Mn^{2+} ions:

$$2Mn^{3+} + C_2O_4^{2-} \rightarrow 2CO_2 + 2Mn^{2+}$$

So, until some Mn^{2+} ions have been formed, the manganate(VII) solution is decolourised only slowly. Such catalysis by a product of the same reaction is called **autocatalysis**.

Other applications of transition-metal complexes

The previous section has shown that it is the availability of variable oxidation states which makes an effective catalyst; this is why transition metals and their compounds find such widespread use in catalysis.

Essential Notes

See the outline of haem representation on page 56.

As well as uses in catalysis, in industrial processes, and in chemical analysis, transition-metal complexes also play a vital, if less obvious, role in human biology.

Our bodies contain an iron(II) complex called **haemoglobin**; this complex is responsible for the red colour of blood and for the transport of oxygen by red blood cells from one part of the body to another. The iron(II) ion in haemoglobin is octahedrally co-ordinated, with the porphyrin ligand providing four nitrogen donor atoms in a square plane around the iron; globin occupies a fifth position and the sixth position is occupied either by water or by molecular oxygen. The affinity of this complex for carbon monoxide is much greater than its affinity for oxygen, and it is the formation of very stable **carboxyhaemoglobin**, inhibiting the uptake of oxygen, that makes carbon monoxide so dangerous.

In medicine too, transition-metal complexes play important roles in drug formulation and use. Perhaps one of the best known examples is the anti-cancer drug **cisplatin**. This drug has had a remarkable success rate in the cure of certain types of cancer, notably those of the testes and ovary; its medicinal use is in chemotherapy. Cisplatin works by forming bonds with guanine bases in DNA, which cause cross-linking between DNA strands. Unwinding of the strands and replication is prevented, so that cancerous cells cannot divide and grow. Unfortunately cisplatin also attacks some healthy cells so that there can be unwanted side effects, the most obvious of which is hair loss. Research continues to develop derivatives of cisplatin (in which ammonia is replaced by other ligands) in the hope that these side effects can be eliminated. Cisplatin, whose structure is shown in Fig 14, is a four-coordinate, square-planar complex of platinum(II), with Cl^- ions and NH_3 molecules as ligands.

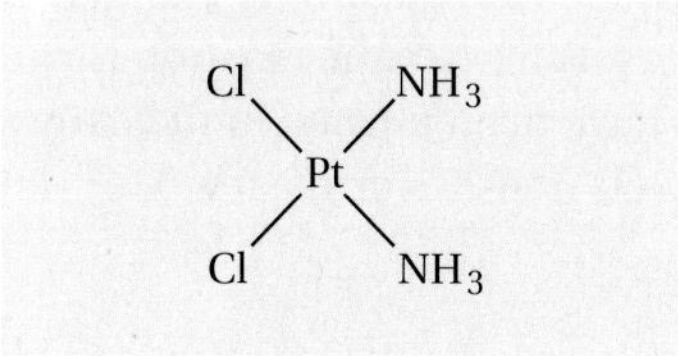

Fig 14
The structure of cisplatin

In section 3.5.4 of this unit it was noted that silver(I) ions usually form linear complexes. The complex ion $[Ag(NH_3)_2]^+$ is formed when silver chloride or silver oxide dissolves in aqueous ammonia to form the colourless solution known as Tollens' reagent. This reagent is used to distinguish aldehydes from ketones; aldehydes give a silver mirror on the side of the test tube, ketones do not (see *Collins Student Support Materials: Unit 4 – Kinetics, Equilibria and Organic Chemistry*, section 3.4.5):

$$RCHO + 2[Ag(NH_3)_2]^+ + 3OH^- \rightarrow RCOO^- + 2Ag + 4NH_3 + 2H_2O$$

3.5.5 Reactions of inorganic compounds in aqueous solution

Lewis acids and bases

In 1938, G.N. Lewis proposed a definition of acids and bases which has become an important concept in chemistry:

> **Definition**
> A **Lewis acid** is an **electron-pair acceptor**.
> A **Lewis base** is an **electron-pair donor**.

Molecules or ions with lone pairs (i.e. non-bonded pairs) of electrons are capable of behaving as bases, and molecules or ions which can accept a pair of electrons are capable of behaving as acids. For example, in the reaction between H_2O and H^+:

H_2O has two *lone pairs* of electrons on the oxygen atom, one of which it can *donate*.

H_2O can act as a *Lewis base*.

H^+ has *no electrons*, so it can *accept* an electron pair (to achieve a full shell).

H^+ can act as a *Lewis acid*.

Consequently, the reaction which occurs between H_2O and H^+ is:

$$H_2O + H^+ \rightarrow H_3O^+$$

Another example is the reaction of boron trifluoride with ammonia:

$$BF_3 + NH_3 \rightarrow [F_3B \leftarrow NH_3]$$

Example

Try to work out which is the acid and which is the base in the reaction between BF_3 and NH_3.

Method

Inspect the equation:

$$BF_3 + NH_3 \rightarrow [F_3B \leftarrow NH_3]$$

to find which species is acting as a *lone-pair donor*. This species is the *Lewis base*.

Answer

Ammonia has a lone pair of electrons on the nitrogen atom, and so is acting as a base. Boron in boron trifluoride has only six electrons in its outer energy level and is capable of accepting two more electrons to give a full complement of eight; it can therefore act as an acid.

Comment

When a Lewis acid reacts with a Lewis base, a co-ordinate bond is formed. A straight arrow is drawn to show the origin of the electron pair in the newly formed co-ordinate bond.

Examiners' Notes

Because it is clear that one electron pair on nitrogen has been donated to boron, it can be helpful to show this process by means of a straight arrow.

In the case of H_3O^+, however, the oxygen atom shares three electron pairs with three *equivalent* hydrogen atoms, so an arrow cannot be drawn to single out any one of them; all three O—H bonds are identical, unlike the B—N bond, which is clearly the only *donor–acceptor* bond in the molecule.

Examiners' Notes

Note the similarity of the straight arrow to the curly arrow used in organic chemistry. The straight arrow also starts at a pair of electrons and points to where they end up in the new species.

Remember:
Lewis base = nucleophile
= ligand

In transition-metal chemistry, all ligands are *Lewis bases* and all metal cations (having a positive charge) are *Lewis acids*. The ligand *donates a pair of electrons* to the metal.

Metal-aqua ions

When a transition-metal ion is placed in water, one of the two lone pairs of electrons on each of six water molecules is used to form a co-ordinate bond with the metal ion. The result is a **co-ordination compound** or **complex**. This reaction and its associated colour change was illustrated in section 3.5.4, using copper(II) sulphate.

A still more striking colour change occurs when anhydrous cobalt(II) chloride is added to water. The anhydrous salt is blue, whereas the hexaaquacobalt(II) cation is pink.

The colours of solutions of transition-metal salts in water are largely due to the colours of the metal-aqua ions, for example:

$[Fe(H_2O)_6]^{2+}$ green

$[Co(H_2O)_6]^{2+}$ pink

$[Cu(H_2O)_6]^{2+}$ blue

Any copper(II) salt in aqueous solution will be present as the hexaaqua ion and the solution will be blue. Anhydrous copper(II) salts are not blue: $CuSO_4$ is white, $CuCl_2$ is yellow, $CuBr_2$ is black, and CuO is also black, yet they all give blue solutions when dissolved in water or dissolved in dilute acids.

So strong is the bonding between the water molecules and the metal ions, that the hexaaqua ion remains intact when solutions containing this ion are evaporated. Thus, solid transition-metal salts in the laboratory are often not simple salts but contain a complex, the metal-hexaaqua ion. Examples of 'salts' containing the hexaaqua ion in the solid state include:

$FeSO_4.7H_2O$	green	contains $[Fe(H_2O)_6]^{2+}$
$CoCl_2.6H_2O$	pink	contains $[Co(H_2O)_6]^{2+}$
$Fe(NO_3)_3.9H_2O$	pale violet	contains $[Fe(H_2O)_6]^{3+}$

Metal(III)-aqua ions

Whilst the reactions of metal(II) salts with water do not appear to be violent, enough energy is given out on hydration to overcome the lattice energy of the solid so that the solid dissolves; reaction with water breaks down the crystal lattice. When the charge on the metal ion is increased to 3+, however, the hydration reaction is often very exothermic. Thus, aluminium chloride dissolves in water violently with evolution of heat and fumes:

$$AlCl_3(s) \xrightarrow{\text{excess } H_2O(l)} [Al(H_2O)_6]^{3+}(aq) + 3Cl^-(aq)$$

Note that there is no precipitate of a metal hydroxide in this reaction; in general, when simply added to water, metal(III) chlorides do not hydrolyse to form metal(III) hydroxides.

Metal ions exist in aqueous solution as 6-water or hexaaqua ions. The hexaaqua ion contains a metal ion surrounded octahedrally by six oxygen atoms from six water molecules. When metal ions react in aqueous solution, it is essential to remember these six water molecules because they can take part in the chemical reactions of metal ions.

When chemical reactions occur, bonds are frequently broken and, more often than not, new bonds are formed. The driving force for reaction is often bond formation. When considering reactions in water, the questions to ask are, 'which bonds can be broken?' and 'which bonds can be formed?'

The bonding in a hexaaqua ion, $[M(H_2O)_6]^{n+}$, is shown in Fig 15.

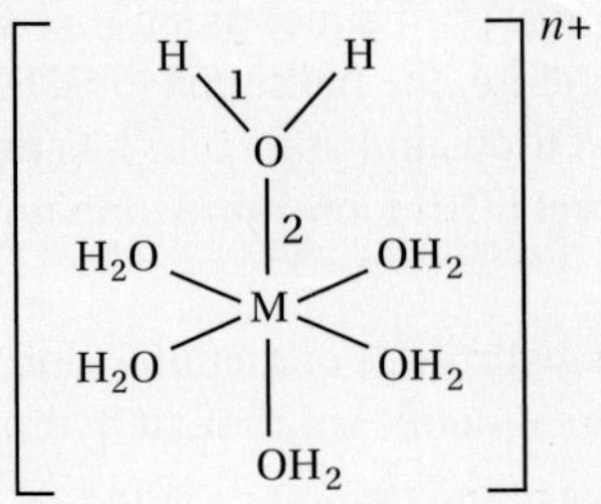

Fig 15
The hexaaqua ion $[M(H_2O)_6]^{n+}$ contains only *two* types of bond:

O—H bonds, covalent bonds in water molecules co-ordinated to the metal ion; one of these bonds is labelled 1.

M—O bonds, co-ordinate bonds between water ligands and the metal ion; one of these bonds is labelled 2.

In reactions of metal ions, the only process which occurs without breaking either of these two types of bond is a redox reaction, i.e. oxidation or reduction, which involves only the loss or gain of an electron, for example:

$$[M(H_2O)_6]^{2+} \rightarrow [M(H_2O)_6]^{3+} + e^-$$

If the O—H bond in a co-ordinated water molecule is broken, the reaction is called an **acidity reaction** or a **hydrolysis reaction**; if the M—O bond is broken (and a new ligand is attached to the metal) the reaction is called a **substitution reaction**. Both of these important reactions are considered in more detail below.

Acidity or hydrolysis reaction

When a metal-aqua ion is placed in water, an equilibrium is set up:

$$[M(H_2O)_6]^{2+} + H_2O \rightleftharpoons [M(H_2O)_5(OH)]^+ + H_3O^+$$

which is called the *hydrolysis* reaction, because a water molecule has been split into OH^- and H^+. It is also called the *acidity* reaction, because hydrolysis leads to the formation of H_3O^+ ions.

For a metal(II) ion, as above, the equilibrium lies very much to the left-hand side, so that metal(II) ions are only slightly acidic in water. The degree of acidity is measured by the magnitude of the equilibrium constant for the reaction (see *Collins Student Support Materials: Unit 4 – Kinetics, Equilibria and Organic Chemistry*, section 3.4.3). For the above equilibrium, the **acid dissociation constant** (or **acidity constant**) is defined as:

$$K_a = \frac{[M(H_2O)_5(OH)^+][H_3O^+]}{[[M(H_2O)_6]^{2+}]}$$

Examiners' Notes

Hydrolysis is a reaction of water in which an O–H bond is broken. A reaction of water in which the water molecule remains intact is called hydration.

Essential Notes

In this relationship, the square brackets indicate the concentrations of the different species.

For metal(II) ions, K_a varies between 10^{-6} and 10^{-11} and **pK_a** varies between 6 and 11.

For metal(III) ions, however, the equilibrium:

$$[M(H_2O)_6]^{3+} + H_2O \rightleftharpoons [M(H_2O)_5(OH)]^{2+} + H_3O^+$$

lies further to the right and, for this equilibrium, K_a varies between 10^{-2} and 10^{-5}, so that pK_a takes values between 2 and 5.

The acidity of transition-metal ions at usual concentrations can be summarised as follows:

$[M(H_2O)_6]^{2+}$ $[M(H_2O)_6]^{3+}$

very weak acids   weak acids

pH around 6   pH around 3

The principal species present in aqueous solutions of metal(II) and metal(III) salts are, therefore, the hexaaqua ions. For metal(II) ions, only about one in a hundred thousand aqua ions has undergone the hydrolysis reaction, whereas for metal(III) ions, more than one in a thousand has been hydrolysed.

Because of the extreme hydrolysis of metal(IV) ions, the ion $[M(H_2O)_6]^{4+}$ does not usually exist in aqueous solution; it hydrolyses to form $M(OH)_4$.

Two factors are involved in deciding the acidity of metal ions; these are:

- the charge on the metal ion   *acidity increases with charge*
- the size of the metal ion   *acidity decreases as size increases*

Small, highly-charged cations are the strongest acids in aqueous solution. If the charge and size factors are taken together, the **charge-to-size ratio** can be used to predict the relative acidities of metal ions. This ratio reflects the **polarising power** of the metal ion. The greater the power of the metal ion to attract electron density from the oxygen atom of a co-ordinated water molecule, the weaker is the O–H bond in this molecule. As a result, it is easier to break this O–H bond and release an H^+ ion to a water molecule which is not co-ordinated.

So far, only the first stage of the hydrolysis reaction has been considered. Further hydrolysis can occur and this is especially true if a base is added to upset the equilibrium.

For metal(II) ions the complete hydrolysis is shown by the equations:

$$[M(H_2O)_6]^{2+} + H_2O \rightleftharpoons [M(H_2O)_5(OH)]^+ + H_3O^+$$

$$[M(H_2O)_5(OH)]^+ + H_2O \rightleftharpoons [M(H_2O)_4(OH)_2] + H_3O^+$$

while for metal(III) ions, the equations are:

$$[M(H_2O)_6]^{3+} + H_2O \rightleftharpoons [M(H_2O)_5(OH)]^{2+} + H_3O^+$$

$$[M(H_2O)_5(OH)]^{2+} + H_2O \rightleftharpoons [M(H_2O)_4(OH)_2]^+ + H_3O^+$$

$$[M(H_2O)_4(OH)_2]^+ + H_2O \rightleftharpoons [M(H_2O)_3(OH)_3] + H_3O^+$$

Note that the final product in all cases is the neutral metal hydroxide. This species has no overall charge and there is therefore little hydration energy available to overcome the lattice energy, so that such hydroxides always appear as precipitates.

● All transition-metal hydroxides are insoluble in water; they are only formed from aquated metal(II) and metal(III) ions when a base (e.g. NaOH or NH_3) is added.

Some reactions, which can be understood only by using these hydrolysis equations, will now be considered. If either some pale violet crystals of iron(III) nitrate nonahydrate, $Fe(NO_3)_3.9H_2O$, or some crystals of iron(III) ammonium alum, $Fe_2(SO_4)_3.(NH_4)_2SO_4.24H_2O$, are added to water, a brown solution forms. Both of these iron salts contain the very pale violet hexaaqua ion $[Fe(H_2O)_6]^{3+}$, and it is clearly this ion which is reacting with the water:

$$[Fe(H_2O)_6]^{3+} + H_2O \rightleftharpoons [Fe(H_2O)_5(OH)]^{2+} + H_3O^+$$
$$\text{pale violet} \qquad\qquad \text{brown}$$

The hydrolysis product is brown and, since this ion is more intensely coloured than the hexaaqua ion, the colour of the solution is seen to be brown. Don't forget, however, that the above equilibrium still lies very far to the left, so that $[Fe(H_2O)_6]^{3+}$ remains the principal species present.

Applying the principles of equilibrium (see *Collins Student Support Materials: Unit 2 – Chemistry in Action,* section 3.2.3) to this case allows prediction of the outcome when either acid or base is added to the system above.

If a little nitric acid is added to the brown solution, the brown colour disappears and an almost colourless (pale violet) solution results. Clearly, if an acid is added to the equilibrium system above, the equilibrium will respond by moving to the left to remove the excess of acid, and some $[Fe(H_2O)_5(OH)]^{2+}$ will be converted into $[Fe(H_2O)_6]^{3+}$. Similarly, if a base such as sodium hydroxide is added to the equilibrium system above, the base will react with H_3O^+ ions, removing some of them; the system will respond by producing more H_3O^+ ions and the equilibrium will move to the right. What is observed on adding acid, therefore, is a decrease of the dark brown colour. On adding base, the solution turns darker brown until eventually (when the next two stages of the hydrolytic equilibria have been moved to the right) a brown precipitate of iron(III) hydroxide forms.

Reactions of aqua ions with alkalis

When sodium hydroxide solution is added to a solution of a transition-metal salt, a precipitate of the metal hydroxide always appears.

For instance, consider what happens when sodium hydroxide solution is added to a solution of copper(II) sulfate. The species present in a solution of copper(II) sulfate are:

$[Cu(H_2O)_6]^{2+}$ (major species), $[Cu(H_2O)_5(OH)]^+$ and H_3O^+ (small amount)

When OH^- ions are added to this mixture, they attack the strongest acid present, which is H_3O^+:

$$OH^- + H_3O^+ \rightarrow 2H_2O$$

This reaction upsets the hydrolysis equilibrium:

$$[Cu(H_2O)_6]^{2+} + H_2O \rightleftharpoons [Cu(H_2O)_5(OH)]^+ + H_3O^+$$

Examiners' Notes

Metal hydroxides can be written as $M(OH)_2$ and $M(OH)_3$. Whilst the hydroxides $M(OH)_2$ do actually occur with this precise formula in the solid state, metal(III) hydroxides readily lose water on drying and usually occur in the solid state as MO(OH). The compound with the formula FeO(OH), for example, is '*rust*'.

and, when this equilibrium has moved completely over to the right-hand side, the next equilibrium is set up:

$$[Cu(H_2O)_5(OH)]^+ + H_2O \rightleftharpoons [Cu(H_2O)_4(OH)_2] + H_3O^+$$

As more sodium hydroxide is added, further H_3O^+ is removed and blue copper(II) hydroxide precipitates. Thus, the overall equation is:

$$[Cu(H_2O)_6]^{2+} + 2OH^- \rightleftharpoons [Cu(OH)_2(H_2O)_4] + 2H_2O$$

The same sequence occurs whatever the metal ion, so that the reactions of sodium hydroxide with transition-metal ions can be summarised as:

● The addition of *sodium hydroxide* to a solution of a transition-metal salt *always precipitates* the transition-metal hydroxide.

The colours of some common hydroxides formed in these reactions are:

$[Fe(H_2O)_6]^{2+}$ green solution	$\rightarrow$	$[Fe(H_2O)_4(OH)_2]$ green precipitate
$[Co(H_2O)_6]^{2+}$ pink solution	$\rightarrow$	$[Co(H_2O)_4(OH)_2]$ blue-green precipitate
$[Fe(H_2O)_6]^{3+}$ pale violet solution	$\rightarrow$	$[Fe(H_2O)_3(OH)_3]$ brown precipitate
$[Cr(H_2O)_6]^{3+}$ red-violet solution	$\rightarrow$	$[Cr(H_2O)_3(OH)_3]$ green precipitate

The addition of aqueous ammonia to a solution of a transition-metal salt results in exactly the same initial sequence of events as occur on the addition of sodium hydroxide. The ammonia upsets the hydrolysis equilibrium by removing H_3O^+ until the metal hydroxide is formed:

$$NH_3 + H_3O^+ \rightarrow NH_4^+ + H_2O$$

● The addition of *ammonia solution* to a solution of a transition-metal salt *always* results, *initially*, in the formation of a *precipitate* of the metal hydroxide.

A further reaction (i.e. *substitution*; see pages 76–83) can occur in the presence of an excess of ammonia, but initially a *precipitate* of the metal hydroxide is always formed.

Reactions of aqua ions with carbonate ions

Sodium carbonate is commonly used as a base, so it is important to be aware of its reactions with metal ions. The carbonate ion reacts with acids to form the hydrogencarbonate ion and then carbonic acid (H_2CO_3), which rapidly becomes carbon dioxide and water:

$$CO_3^{2-} + H_3O^+ \rightleftharpoons HCO_3^- + H_2O$$

$$HCO_3^- + H_3O^+ \rightleftharpoons CO_2 + 2H_2O$$

Metal(II) ions are not sufficiently acidic to displace carbonic acid from carbonates. Instead, these ions form insoluble metal carbonates, $MCO_3(s)$.

● The addition of *sodium carbonate solution* to a solution of a metal(II) salt results in a *precipitate* of the metal(II) carbonate.

Metal(III) ions, however, are more acidic than carbonic acid and therefore displace this acid from solutions containing carbonate ions (a strong acid will always displace a weaker one). So the reaction between a solution of metal(III) ions and carbonate ions results in *effervescence* of carbon dioxide and *precipitation* of the metal hydroxide.

The reaction follows the scheme:

$$[M(H_2O)_6]^{3+} + H_2O \rightleftharpoons [M(H_2O)_5(OH)]^{2+} + H_3O^+$$

followed by the reaction of H_3O^+ with CO_3^{2-}:

$$2H_3O^+ + CO_3^{2-} \rightleftharpoons CO_2 + 3H_2O$$

so that the equilibrium is pushed to the right and subsequent equilibria are set up:

$$[M(H_2O)_5(OH)]^{2+} + H_2O \rightleftharpoons [M(H_2O)_4(OH)_2]^+ + H_3O^+$$

$$[M(H_2O)_4(OH)_2]^+ + H_2O \rightleftharpoons [M(H_2O)_3(OH)_3] + H_3O^+$$

until the metal(III) hydroxide precipitates.

In general:

- The reaction between a metal(III) salt solution and sodium carbonate leads to *evolution of a gas* (carbon dioxide) and *precipitation* of the metal(III) hydroxide.

As a result, metal(III) carbonates cannot be prepared in aqueous solution and, in general, they do not exist. So, for example, $Al_2(CO_3)_3$, $Fe_2(CO_3)_3$ and $Cr_2(CO_3)_3$ *cannot* be prepared in aqueous solution and are, indeed, unknown.

The three equations in the hydrolysis of a metal(III) ion can be added together to give the overall equation:

$$[M(H_2O)_6]^{3+} + 3H_2O \rightleftharpoons [M(H_2O)_3(OH)_3] + 3H_3O^+$$

This equilibrium is driven to the right following removal of H_3O^+ by carbonate ions:

$$2H_3O^+ + CO_3^{2-} \rightarrow CO_2 + 3H_2O$$

Combining these two equations gives:

$$2[M(H_2O)_6]^{3+} + 3CO_3^{2-} \rightarrow 2[M(H_2O)_3(OH)_3] + 3CO_2 + 3H_2O$$

Amphoteric character

All of the hydrolysis equilibria given above can be reversed by using strong acids. Thus, metal hydroxides dissolve in strong acids to give metal-aqua ions (as long as a non-complexing strong acid such as HNO_3 is used; nitrate ions are only *weak* ligands).

Whilst metal hydroxides do not undergo hydrolysis in water, they can be attacked by the strong nucleophile OH^- to give anionic complexes which are water-soluble. This ability of metal hydroxides to dissolve in both strong acids and strong alkalis is known as *amphoteric character* (see section 3.5.2).

When sodium hydroxide is added to a solution of an aluminium salt, a white precipitate of aluminium hydroxide is seen initially. Then, as an excess of the alkali is added, the precipitate dissolves to form a colourless solution containing the aluminate ion:

$$[Al(H_2O)_6]^{3+} + 3H_2O \rightleftharpoons [Al(H_2O)_3(OH)_3] + 3H_3O^+$$

$$[Al(H_2O)_3(OH)_3] + OH^- \rightleftharpoons [Al(OH)_4]^- + 3H_2O$$

Examiners' Notes

The effervescence is best observed using solid sodium carbonate. In dilute aqueous solutions, the carbon dioxide is absorbed by the aqueous carbonate ions to form hydrogencarbonate ions:

$$CO_3^{2-} + H_2O + CO_2 \rightarrow 2HCO_3^-$$

Examiners' Notes

It is not only carbonate ions which behave in this way. Anions derived from weak acids are protonated by metal(III)-aqua ions so that the free acid and the metal(III) hydroxide result.

Some examples of anions of weak acids are S^{2-}, NO_2^-, SO_3^{2-}, $S_2O_3^{2-}$ and CH_3COO^-. Metal(III) salts of these anions cannot usually be prepared in aqueous solution.

Examiners' Notes

Whereas *all bases* yield hydroxide precipitates, only *strong bases in excess* can cause some $M(H_2O)_3(OH)_3$ precipitates to re-dissolve, forming negatively-charged complexes. An excess of a *weak base* does not re-dissolve these hydroxide precipitates.

Similarly, the addition of sodium hydroxide to a solution of a chromium(III) salt leads to the initial formation of a green precipitate of chromium(III) hydroxide, which then dissolves to give a green solution containing chromate(III) ions when an excess of the alkali is added:

$$[Cr(H_2O)_6]^{3+} + 3H_2O \rightleftharpoons [Cr(H_2O)_3(OH)_3] + 3H_3O^+$$

$$[Cr(H_2O)_3(OH)_3] + 3OH^- \rightleftharpoons [Cr(OH)_6]^{3-} + 3H_2O$$

The general scheme for metal(III) ions is:

$$[M(H_2O)_6]^{3+} \underset{H_3O^+}{\overset{OH^-}{\rightleftharpoons}} [M(H_2O)_3(OH)_3] \underset{H_3O^+}{\overset{OH^-}{\rightleftharpoons}} [M(OH)_4]^-$$

acidic solution   neutral solution   alkaline solution

Most transition-metal hydroxides are amphoteric.

In high oxidation states, transition metals often exist in solution in an anionic form. An important anionic equilibrium occurs in chromium(VI) chemistry. Chromate(VI) ions (see section 3.5.4) react with acids to form dichromate(VI) ions:

$$2CrO_4^{2-} + 2H^+ \rightleftharpoons Cr_2O_7^{2-} + H_2O$$

yellow   orange

chromate(VI)   dichromate(VI)

The dichromate(VI) ion is converted back into the chromate(VI) ion in alkaline solution:

$$Cr_2O_7^{2-} + 2OH^- \rightleftharpoons 2CrO_4^{2-} + H_2O$$

Summaries of the reactions of metal(II) and metal(III) ions with bases are given in Tables 23 and 24 on pages 78 and 79.

Substitution reactions

Each of the acidity reactions above has involved the breaking of an O–H bond in a co-ordinated water molecule. Other reactions in which one or more of the M–O bonds in the hexaaqua ion are broken will now be considered. When a water molecule is replaced by another ligand, a substitution reaction occurs; it is *nucleophilic substitution*, but is often just called **ligand substitution**.

For the purposes of substitution reactions, there are two types of ligand:

- *neutral* – uncharged molecules
- *anionic* – negatively-charged ligands.

Substitution by neutral ligands

Ammonia can act as a base in the *Brønsted–Lowry* sense (when it reacts with a proton, as it does in the acidity reaction considered above) and also as a base in the *Lewis* sense (when it acts as a ligand).

A general equation can be written for the replacement of water molecules in an aqua ion by neutral ligands. For example, the replacement of H_2O by NH_3 can be written as:

$$[M(H_2O)_6]^{2+} + 6NH_3 \rightleftharpoons [M(NH_3)_6]^{2+} + 6H_2O$$

The equation above disguises the fact that what is written as a one-step equilibrium can be broken down into six steps, with only one water molecule being replaced in each step:

$$[M(H_2O)_6]^{2+} + NH_3 \rightleftharpoons [M(NH_3)(H_2O)_5]^{2+} + H_2O$$

$$[M(NH_3)(H_2O)_5]^{2+} + NH_3 \rightleftharpoons [M(NH_3)_2(H_2O)_4]^{2+} + H_2O$$

$$[M(NH_3)_2(H_2O)_4]^{2+} + NH_3 \rightleftharpoons [M(NH_3)_3(H_2O)_3]^{2+} + H_2O$$

$$[M(NH_3)_3(H_2O)_3]^{2+} + NH_3 \rightleftharpoons [M(NH_3)_4(H_2O)_2]^{2+} + H_2O$$

$$[M(NH_3)_4(H_2O)_2]^{2+} + NH_3 \rightleftharpoons [M(NH_3)_5(H_2O)]^{2+} + H_2O$$

$$[M(NH_3)_5(H_2O)]^{2+} + NH_3 \rightleftharpoons [M(NH_3)_6]^{2+} + H_2O$$

At first sight, these equations look complicated, but they are in fact quite simple. The first equation starts with a hexaaqua ion and the last equation ends with a hexaammine ion. Note that, when water is bonded to a metal ion, it is called *aqua*, but when ammonia is bonded, it is called *ammine*.

Ammonia is uncharged and has a similar size to water. Consequently, no change of shape is expected to occur during these substitution reactions. Thus, all the complexes in the equilibria above are *octahedral*.

One of the most famous historical examples of such a substitution sequence occurs in the addition of ammonia to cobalt(II) ions. In this sequence, the first change observed is the formation of a *green-blue precipitate*. This precipitate is cobalt(II) hydroxide, formed by the acidity reaction of the cobalt(II) ions. The overall reaction for this process can be written as:

$$[Co(H_2O)_6]^{2+} + 2NH_3 \rightleftharpoons [Co(H_2O)_4(OH)_2] + 2NH_4^+$$
pink solution   green-blue precipitate

When an excess of concentrated aqueous ammonia is added to the reaction mixture, the green-blue precipitate dissolves and a *pale straw-coloured solution* results. It is important to keep air away from this solution, which otherwise darkens rapidly to form a *dark-brown mixture* containing cobalt(III) ammines.

The species in the straw-coloured solution is the hexaamminecobalt(II) ion. The green-blue hydroxide dissolves in ammonia according to the equation:

$$[Co(H_2O)_4(OH)_2] + 6NH_3 \rightleftharpoons [Co(NH_3)_6]^{2+} + 4H_2O + 2OH^-$$
green-blue precipiate   straw-coloured solution

The overall equation, starting from the aqua ion, is:

$$[Co(H_2O)_6]^{2+} + 6NH_3 \rightleftharpoons [Co(NH_3)_6]^{2+} + 6H_2O$$
pink solution   straw-coloured solution

Thus, with cobalt(II), complete substitution of water molecules by ammonia molecules occurs.

With copper(II) ions, however, only four of the six water molecules on copper are replaced. As with cobalt, when aqueous ammonia is added to a solution

Essential Notes

Note the double 'm' in *ammine*; an *amine* is quite different.

of a copper(II) salt, the first change observed is the formation of a *blue precipitate* of the hydroxide. This then dissolves when an excess of ammonia is added and a *deep-blue solution* of the tetraamminebisaquacopper(II) ion is formed.

$$[Cu(H_2O)_6]^{2+} + 4NH_3 \rightleftharpoons [Cu(NH_3)_4(H_2O)_2]^{2+} + 4H_2O$$

blue solution   deep-blue solution

Further substitution can be achieved when the concentration of ammonia is increased (e.g. by cooling the solution in ice and saturating it with ammonia gas, or by using liquid ammonia rather than aqueous ammonia), but in concentrated aqueous ammonia, the equilibrium position reached is that in which the dark-blue $[Cu(NH_3)_4(H_2O)_2]^{2+}$ ion is formed. This ion has four ammonia molecules in a square-planar arrangement around copper with water molecules occupying the other two octahedral positions above and below the plane. The bonds to water are longer and weaker than the bonds to ammonia.

A summary of the reactions of metal(II) and metal(III) ions with various bases (hydroxide ions, ammonia and carbonate ions) is given in Tables 23 and 24.

Examiners' Notes

A more sophisticated explanation for the lack of formation of the hexaammine, based on the Jahn–Teller effect, can be found in undergraduate texts.

Table 23
Summary of the reactions of metal(II) ions with bases

Base added	Aqueous M(II) ions		
	$[Fe(H_2O)_6]^{2+}$ green solution	$[Co(H_2O)_6]^{2+}$ pink solution	$[Cu(H_2O)_6]^{2+}$ blue solution
OH⁻ (little)	$[Fe(H_2O)_4(OH)_2]$ green ppt	$[Co(H_2O)_4(OH)_2]$ blue ppt	$[Cu(H_2O)_4(OH)_2]$ blue ppt
OH⁻ (excess dilute) NH₃ (little)	(turns brown in air)	(turns pink on standing and then turns brown in air)	
NH₃ (excess)	as above	$[Co(NH_3)_6]^{2+}$ pale-brown (straw) solution (turns brown in air)	$[Cu(NH_3)_4(H_2O)_2]^{2+}$ deep-blue solution
CO_3^{2-}	FeCO₃ green ppt	CoCO₃ pink ppt	CuCO₃ green-blue ppt

Base added	Aqueous M(III) ions		
	$[Fe(H_2O)_6]^{3+}$ violet solution (appears yellow due to hydrolysis)	$[Al(H_2O)_6]^{3+}$ colourless solution	$[Cr(H_2O)_6]^{3+}$ dull red-blue solution (ruby)
OH^- (little)	$[Fe(H_2O)_3(OH)_3]$ brown ppt	$[Al(H_2O)_3(OH)_3]$ white ppt	$[Cr(H_2O)_3(OH)_3]$ green ppt
OH^- (excess)	as above brown ppt	$[Al(OH)_4]^-$ colourless solution	$[Cr(OH)_6]^{3-}$ green solution
NH_3 (little)	as above brown ppt	$[Al(H_2O)_3(OH)_3]$ white ppt	$[Cr(H_2O)_3(OH)_3]$ green ppt
NH_3 (excess)	as above brown ppt	as above white ppt	$[Cr(NH_3)_6]^{3+}$ purple solution
CO_3^{2-}	as above brown ppt and effervescence of CO_2	as above white ppt and effervescence of CO_2	$[Cr(H_2O)_3(OH)_3]$ green ppt and effervescence of CO_2

Table 24
Summary of the reactions of metal(III) ions with bases

Substitution by chloride ions

All anions are potentially capable of acting as ligands. For example, the very common ligand Cl^- has the electron arrangement of the noble gas argon; it has four lone pairs of electrons. When Cl^- bonds to a metal ion, one of these lone pairs forms a co-ordinate bond to the metal; the three remaining lone pairs on the chloride ion do not co-ordinate.

A good source of chloride ions is concentrated hydrochloric acid. Hydrogen chloride is very much more soluble in water than ionic chlorides such as sodium chloride, allowing high concentrations to be achieved; this is why concentrated hydrochloric acid is used to prepare complexes of chloride ions with transition metals.

When a pink solution of a cobalt(II) salt in water is treated with an excess of concentrated hydrochloric acid, a deep-blue solution is formed:

$$[Co(H_2O)_6]^{2+} + 4Cl^- \rightleftharpoons [CoCl_4]^{2-} + 6H_2O$$

pink   blue

octahedral   tetrahedral

When the blue solution of $[CoCl_4]^{2-}$ is diluted with water, the pink colour returns. This behaviour can be understood in terms of an equilibrium which can be driven from left to right by high chloride ion concentrations and from right to left if the concentration of chloride ions is lowered by dilution. This marked change in colour is brought about not only by the change of ligand but, more significantly, also by the change in the co-ordination number of the cobalt ion. The shape of the ligands around the cobalt ion changes from octahedral to tetrahedral as hydrochloric acid is added.

Why then does the shape of the complex ion change? The chloride ligand is:

- negatively charged
- and large.

As the size of the ligands around a metal ion is increased, a point is reached when the electron charge-clouds around the ligands repel each other to such an extent that the octahedral structure becomes less stable than the tetrahedral one. In the tetrahedral structure, the ligands are approximately 109° apart and do not experience repulsive forces as great as in the octahedral arrangement, where the ligands are only 90° apart.

Thus, the general equation for the substitution reaction of a metal(II)-aqua ion with chloride ions is:

$$[M(H_2O)_6]^{2+} + 4Cl^- \rightleftharpoons [MCl_4]^{2-} + 6H_2O$$

octahedral   tetrahedral

Examiners' Notes

If the size of the ligand is decreased to the smaller F^- ion, then octahedral complexes become common, e.g. AlF_6^{3-} (c.f. $AlCl_4^-$). If, however, the size of the ligand is increased to the larger I^- ion, octahedral complexes are not formed with any of the cations of the first three periods.

Example

Deduce what will happen when concentrated hydrochloric acid is added to a solution of copper(II) sulfate.

Method

Recall the effect of ligand substitution on shape, co-ordination number, and colour in a complex ion.

Answer

The copper(II)-aqua ion is *octahedral*. When small water molecules are replaced by bigger chloride ions, a *tetrahedral* complex is likely to be formed. The *co-ordination number* will fall from 6 to 4. There will probably be a change of colour in the solution.

Comment

The blue solution of copper(II) ions turns yellow-green once an excess of hydrochloric acid has been added and the tetrachlorocuprate(II) ion is formed:

Equation:	$[Cu(H_2O)_6]^{2+} + 4Cl^- \rightleftharpoons [CuCl_4]^{2-} + 6H_2O$	
Colour:	blue	yellow-green
Shape:	octahedral	tetrahedral
Co-ordination number:	6	4

When the solution is diluted, the decreased concentration of chloride ions forces the equilibrium back to the hexaaqua ion and the solution turns blue again.

Substitution by bidentate and multidentate ligands

When a ligand with two donor atoms attacks a metal-aqua ion, two water molecules are replaced in the first instance. For example, a bidentate ligand (see section 3.5.4) such as ethane-1,2-diamine (*ethylenediamine* or *en*) reacts with a metal(II)-aqua ion according to the equation:

$$[M(H_2O)_6]^{2+} + H_2NCH_2CH_2NH_2 \rightleftharpoons [M(H_2O)_4(H_2NCH_2CH_2NH_2)]^{2+} + 2H_2O$$

As more ligand is added, further substitution can occur until all the water molecules have been replaced:

$$[M(H_2O)_4(H_2NCH_2CH_2NH_2)]^{2+} + H_2NCH_2CH_2NH_2 \rightleftharpoons [M(H_2O)_2(H_2NCH_2CH_2NH_2)_2]^{2+} + 2H_2O$$

$$[M(H_2O)_2(H_2NCH_2CH_2NH_2)_2]^{2+} + H_2NCH_2CH_2NH_2 \rightleftharpoons [M(H_2NCH_2CH_2NH_2)_3]^{2+} + 2H_2O$$

In these reactions with ethylenediamine, the equilibrium position lies well over to the right-hand side; the equilibrium constants for the formation of tris(ethylenediamine) complexes are of the order of 10^{20}. This large value means that the resulting complexes are much more stable than the aqua ions from which they were formed. Note that, because the donor atoms are small, there is no change in shape and the complexes remain octahedral.

Metal(III)-aqua ions react similarly, so that the equation for the reaction with an excess of ethane-1,2-diamine is:

$$[M(H_2O)_6]^{3+} + 3H_2NCH_2CH_2NH_2 \rightleftharpoons [M(H_2NCH_2CH_2NH_2)_3]^{3+} + 6H_2O$$

These metal(III)-tris(ethylenediamine) complexes are even more stable than the metal(II) complexes; the equilibrium constant for the above reaction is of the order of 10^{30}.

This extra stability is known as the **chelate effect**. Complexes in which bidentate or multidentate ligands bond to one metal ion only are known as chelates. The ligand forms a five- or six-membered ring with the metal ion, as shown in Fig 16.

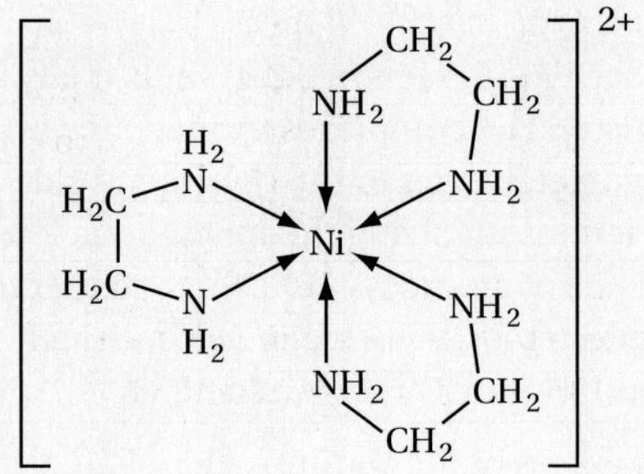

Fig 16
Chelating ethylenediamine

Chelating ligands can be present in complexes which also contain unidentate ligands as in, for example, the ion shown in Fig 17, *trans*-dichlorobis(ethylenediamine)cobalt(III).

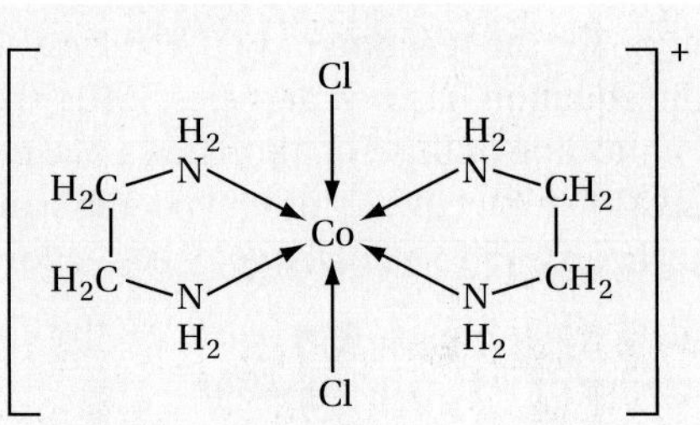

Examiners' Notes

Hint: When drawing complex ions with *multidentate ligands, first* draw the six co-ordinate bonds arranged octahedrally around the central metal atom, *then* fit the ligands to these bonds.

Fig 17
Unidentate chloride ions as part of a bidentate chelate complex

Another common bidentate ligand is the di-anion ethanedioate, $C_2O_4^{2-}$. This ligand forms a very stable complex with iron(III) ions, $[Fe(C_2O_4)_3]^{3-}$, shown in Fig 18.

Fig 18
The $[Fe(C_2O_4)_3]^{3-}$ ion

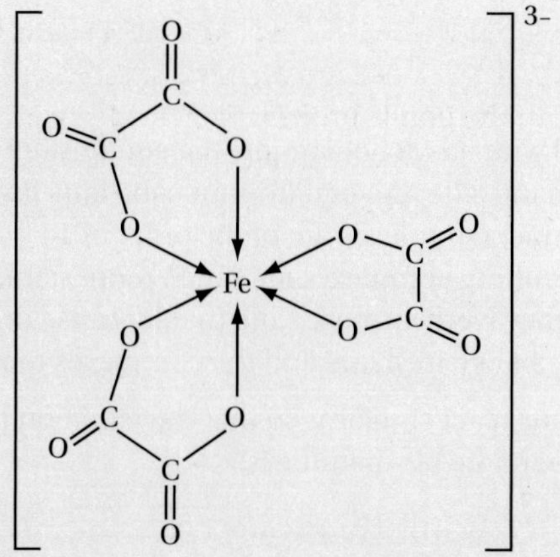

Multidentate ligands form even more stable complexes than do bidentate ligands. Haem in blood (see section 3.5.4) is an example taken from iron chemistry. The **hexadentate** ligand $EDTA^{4-}$ reacts with metal(II)-aqua ions to form a complex in which the ligand occupies all six octahedral sites around the metal ion, liberating six water molecules:

$$[M(H_2O)_6]^{2+} + EDTA^{4-} \rightleftharpoons [M(EDTA)]^{2-} + 6H_2O$$

The $EDTA^{4-}$ ligand (see Fig 10, section 3.5.4) finds many uses in analytical and industrial chemistry. The complexes formed are very stable, so that all the equilibria are completely over to the right-hand side and very few aqua ions are present. Reactions involving the metal-aqua ion are therefore not possible in the presence of an excess of $EDTA^{4-}$, so that metal ions can be kept in solution (sequestered), even when anions which normally cause precipitation, such as OH^- or CO_3^{2-}, are added.

Thus, addition of an excess of the sodium salt of $EDTA^{4-}$ to aqueous copper(II) sulfate, followed by aqueous sodium hydroxide, does not give any precipitate of copper(II) hydroxide because the Cu^{2+} ions have been *sequestered*.

There is a thermodynamic reason for the stability of metal chelates. Consider the equation for the reaction of $EDTA^{4-}$ given above. The left-hand side of the equation has *two* particles, $[M(H_2O)_6]^{2+}$ and $EDTA^{4-}$, whereas the right-hand side of this equation has *seven* particles, $[M(EDTA)]^{2-}$ and six H_2O molecules. This considerable increase in randomness results in an increase in *entropy* (see section 3.5.1).

Recall that a chemical reaction becomes *feasible* if the change in free energy, $\Delta G^{\ominus}$, is negative or zero. The equation:

$$\Delta G^{\ominus} = \Delta H^{\ominus} - T\Delta S^{\ominus}$$

Examiners' Notes

Use of this **sequestering ability** is made in water softening, where the deposition of calcium carbonate in pipes can be prevented.

shows that there will *always* be a negative free-energy change if the enthalpy change, $\Delta H^{\ominus}$, is *negative* and the entropy change, $\Delta S^{\ominus}$, is *positive*. Even when $\Delta H^{\ominus}$ is *positive* (but not very much so), it will readily be outweighed by the $T\Delta S^{\ominus}$ term if the entropy change is *large and positive*.

In the formation of a chelate, $\Delta S^{\ominus}$ is indeed *large and positive* (two particles form seven particles). $\Delta H^{\ominus}$ for such reactions is usually quite small because the bonds formed are the same in number and rather similar in strength to the bonds broken. Even if $\Delta H^{\ominus}$ were to be slightly positive, this term would be outweighed heavily by $-T\Delta S^{\ominus}$. Consequently, $\Delta G^{\ominus}$ is *always very negative* and the reaction is *always feasible*. Reactions of this kind are sometimes described as being entropy driven, because the enthalpy term can usually be ignored.

Examiners' Notes

The increased entropy leads to a large and positive $T\Delta S^{\ominus}$ term which, because it appears in the equation with a negative sign, contributes significantly to ensuring that $\Delta G^{\ominus}$ is very negative.

How Science Works

The introduction of the *How Science Works* component into the new A-level specifications has made formal an approach to the teaching of topics in science which many teachers have, in fact, already been using.

Irrespective of future careers, science students need to become proficient in dealing with the various issues included in *How Science Works* so as to achieve a level of scientific awareness. In order to gain an appreciation of how chemists, in particular, work (using the scientific method) it is necessary to understand and be able to apply the concepts, principles and theories of chemistry. The ways that chemical theories and laws are developed, together with the potential impact of new discoveries on society in general, should become clear to you as new concepts in the specification are explored.

Science starts with experimental observation and investigation, followed by verification (i.e. confirmation by others that the results are reliable). A theory or model is proposed to try to explain a set of observations, which may themselves have been accidental or planned as part of a series of experiments. During A-level chemistry courses, many different types of experiment will be carried out and evaluated.

Once an initial theory, or hypothesis, has been put forward to explain a set of results, further experiments are carried out to test the ability of this theory to make accurate predictions. An initial theory may need to be adapted to take into account fresh evidence. More observations and experimental results are produced and the cycle is continued until a firm theory can be established. It is very important that all experiments are carried out objectively, without any bias towards a desired result.

A-level Chemistry courses provide various opportunities for students to analyse verified experimental data as well as the chance to develop theories based on novel findings. In some instances, it will be recognised that there is insufficient experimental evidence for a firm theory to be accepted. In other cases, different experiments carried out by different people will produce contradictory results. Sometimes an apparently unusual result or observation, if verified, will be of great significance. There then follow opportunities to debate these issues and to understand how conflicting theories can be resolved by the accumulation of further evidence.

The concepts and principles dealt with under *How Science Works* will be assessed during the examination. The questions set will require only a knowledge and understanding of the topics included in the specification. In some instances, however, the ability to analyse and make deductions from unfamiliar information may be required. Typical examination questions are provided at the end of the Unit and aspects of *How Science Works* are highlighted.

Practice exam-style questions

1   The table below gives the enthalpy changes needed to draw a Born–Haber cycle for sodium bromide.

Standard enthalpies		$\Delta H^{\ominus}/\text{kJ mol}^{-1}$
$\Delta H^{\ominus}_{\text{diss}}$	bond dissociation of gaseous bromine molecules	+194
$\Delta H^{\ominus}_{\text{ea}}$	electron affinity of gaseous bromine atoms	−325
$\Delta H^{\ominus}_{\text{f}}$	formation of solid sodium bromide	−361
$\Delta H^{\ominus}_{\text{i}}$	first ionisation of gaseous sodium atoms	+498
$\Delta H^{\ominus}_{\text{sub}}$	sublimation of metallic sodium	+107
$\Delta H^{\ominus}_{\text{L}}$	lattice dissociation of solid sodium bromide	+753
$\Delta H^{\ominus}_{\text{vap}}$	vaporisation of liquid bromine	to be determined

(a)   Using the steps shown in the outline cycle below, complete the Born–Haber cycle for sodium bromide.

On the blue answer lines alongside the enthalpy changes ΔH_1 to ΔH_5, write the appropriate symbol or combination of symbols from the first column in the table above.

On the other four blue lines, write the appropriate balanced chemical equations, including state symbols, for each species.

ion(s)1

ΔH_3 =

ΔH_4 =

atoms

ion(s)2

ΔH_2 =

ΔH_5 =

elements

ΔH_1 =

salt   NaBr(s)

9 marks

(b) **(i)** Use the Born–Haber cycle in part (a) and the data given in the table on page 85 to calculate the enthalpy of vaporisation, $\Delta H^{\ominus}_{vap}$, for 1 mol of liquid bromine.

(ii) The lattice dissociation enthalpy of NaBr calculated using the perfect ionic model has a value of 752 kJ mol^{-1}. What conclusions about the type of bonding in solid NaBr can be drawn from this observation? Explain your answer.

Type of bonding

Explanation

7 marks

(c) **(i)** Explain what is meant by the term *standard enthalpy of hydration*, $\Delta H^{\ominus}_{hyd}$, of an ion.

(ii) Use the data in the table below to determine the standard enthalpy of hydration of the potassium ion.

$Cl^-(g) \rightarrow Cl^-(aq)$	$\Delta H^{\ominus} = -364$ kJ mol^{-1}
$KCl(s) \rightarrow K^+(g) + Cl^-(g)$	$\Delta H^{\ominus} = +718$ kJ mol^{-1}
$KCl(s) \rightarrow K^+(aq) + Cl^-(aq)$	$\Delta H^{\ominus} = +17$ kJ mol^{-1}

6 marks

Total Marks: 22

2 The feasibility of a chemical reaction depends on the standard free-energy change, $\Delta G^{\ominus}$.

(a) Write an equation that relates $\Delta G^{\ominus}$ to the standard enthalpy change, $\Delta H^{\ominus}$, and the standard entropy change, $\Delta S^{\ominus}$.

1 mark

(b) In terms of $\Delta G^{\ominus}$, state the necessary condition for a feasible reaction.

1 mark

(c) At a pressure of 100 kPa, water will not freeze while the surrounding temperature remains above 0 °C.

   (i) State the sign of the enthalpy change and of the entropy change during freezing.

   Sign of $\Delta H^{\ominus}$ _____

   Sign of $\Delta S^{\ominus}$ _____

   (ii) Explain, in terms of ΔG, why water does not freeze at temperatures above 0 °C.

   _____

   _____

   _____ 4 marks

(d) When sodium hydrogencarbonate is added to dilute hydrochloric acid at room temperature, the temperature of the reaction mixture drops. Despite this, the reaction is feasible. Explain why this is so.

   _____

   _____

   _____

   _____

   _____ 4 marks

(e) Potassium chlorate(V) decomposes on heating according to the equation:

$$4KClO_3(s) \rightarrow 3KClO_4(s) + KCl(s) \qquad \Delta H^{\ominus} = +16.8 \text{ kJ mol}^{-1}$$

The table below shows the standard molar entropy, $S^{\ominus}$, for each of the species involved in the reaction.

	$S^{\ominus}/\text{J K}^{-1} \text{ mol}^{-1}$
$KClO_3(s)$	112
$KClO_4(s)$	134
$KCl(s)$	83

   (i) Calculate the standard entropy change for the decomposition of $KClO_3$ according to the equation above.

   _____

   _____

   _____

   _____

   (ii) Calculate the standard free-energy change at 298 K for this decomposition.

   _____

   _____

   _____

   _____

(iii) At what temperature does the thermal decomposition of potassium chlorate(V) become feasible?

_____

_____

_____ 9 marks

Total Marks: 19

3 **(a)** In terms of electron transfer, what is the role of a reducing agent?

_____ 1 mark

(b) Standard electrode potentials, $E^{\ominus}$, are measured relative to a standard reference electrode.

(i) Name the standard reference electrode and state its potential.

Standard reference electrode _____

Electrode potential _____

(ii) State **four** conditions that must apply when values of $E^{\ominus}$ are being measured using this standard reference electrode.

Condition 1 _____

Condition 2 _____

Condition 3 _____

Condition 4 _____

(iii) A *salt bridge* is used when making such measurements. Explain the function of a *salt bridge*, and state what chemical substance it might contain.

Function _____

Contents _____ 8 marks

(c) Explain the term *electrochemical series*.

_____

_____

_____ 2 marks

(d) The standard electrode potentials, $E^{\ominus}$, for two electrodes are shown below. Use this information to answer the questions that follow.

Iron electrode immersed in a solution of aqueous iron(II) ions $\qquad E^{\ominus}/V = -0.44$

Zinc electrode immersed in a solution of aqueous zinc(II) ions $\qquad E^{\ominus}/V = -0.76$

(i) Write the half-equation that applies to the standard electrode potential for zinc.

_____

_____

(ii) The two electrodes and their solutions are used to form an electrochemical cell. Give the conventional cell representation (conventional cell diagram) for the resulting cell and calculate its e.m.f.

Cell representation _____

Cell e.m.f _____

(iii) State which species is reduced when the two electrodes are connected together. Explain your answer.

Species reduced _____

Explanation _____

_____ 7 marks

Total Marks: 18

4 Electrochemical cells can be used as commercial sources of electrical energy.

(a) The lithium cell is classed as *disposable* and has a relatively long life. The materials used are lithium metal and manganese(IV) oxide, both of which are fairly inexpensive. In this cell, the overall reaction is:

$$Li(s) + MnO_2(s) \rightarrow LiMnO_2(s)$$

(i) Explain what, in this context, is meant by the term *disposable*.

_____

(ii) Give **one** benefit and **one** risk to society of the use of such cells.

Benefit _____

_____

Risk _____

_____

(iii) Given that one of the reactions in this cell involves the oxidation of lithium, write the half-equations for the oxidation and reduction reactions in the lithium cell.

Oxidation reaction _____

_____

Reduction reaction _____

_____ 5 marks

(b) The cell reaction in the *dry cell* (Leclanché cell) is:

$$Zn(s) + 2MnO_2(s) + 4NH_4^+(aq) + 2OH^-(aq) \rightarrow [Zn(NH_3)_4]^{2+}(aq) + 2MnO(OH)(s) + 2H_2O(l)$$

This cell has e.m.f. = 1.51 V and a standard enthalpy change for the cell reaction, $\Delta H^\ominus = -263$ kJ mol^{-1}.

Give three changes in conditions that would cause the e.m.f. of this cell to *increase*.

Change 1 _____

Change 2 _____

Change 3 _____ 3 marks

(c) The lead–acid battery is commonly used as a source of electrical energy in motor vehicles. The overall cell reaction and one of the two half-equations in this cell are shown below.

$$PbO_2(s) + Pb(s) + 2H^+(aq) + 2HSO_4^-(aq) \rightleftharpoons 2PbSO_4(s) + 2H_2O(l) \qquad e.m.f. = +2.04 \text{ V}$$

$$PbO_2(s) + 3H^+(aq) + HSO_4^-(aq) + 2e^- \rightarrow PbSO_4(s) + 2H_2O(l) \qquad E^\ominus = +1.68 \text{ V}$$

(i) Explain the significance of the equilibrium arrows in the overall cell equation.

_____

_____

_____

(ii) Deduce the other half-equation in this cell and calculate its standard electrode potential.

Half-equation _____

_____

_____

_____

Value of $E^\ominus$ _____

_____ 6 marks

(d) Commercial hydrogen–oxygen fuel cells are operated with an electrolyte of concentrated aqueous potassium hydroxide at a temperature of 200 °C under a pressure of 3000 kPa.

(i) Write the half-equation for the reduction of water to form aqueous hydroxide ions and one mole of gaseous hydrogen that occurs at one of the porous electrodes in this fuel cell. The standard electrode potential for this process is –0.83 V.

_____

(ii) Write the half-equation for the reaction of one mole of gaseous oxygen with water to form aqueous hydroxide ions at the other porous electrode in this fuel cell. The standard electrode potential for this process is +0.40 V.

_____

(iii) Hence deduce the overall equation in this fuel cell and calculate the *cell e.m.f.*

Overall equation _____

_____

_____

Cell e.m.f. _____

(iv) Justify the conditions of pressure and temperature given on page 90 in the operation of commercial hydrogen–oxygen fuel cells.

Pressure _____

_____

Temperature _____

_____ 9 marks

Total Marks: 23

5 **(a)** Explain, in terms of their structural types, why the melting point of magnesium oxide (MgO) is much higher than that of phosphorus(V) oxide (P_4O_{10}).

Structural type of MgO _____

Structural type of P_4O_{10} _____

Explanation of different melting points _____

_____

_____ 4 marks

(b) Describe what you would see, and write an equation for any reaction occurring, if samples of MgO and P_4O_{10} were to be added separately to water.

Observation with MgO _____

Equation with MgO _____

Observation with P_4O_{10} _____

Equation with P_4O_{10} _____ 4 marks

(c) **(i)** Give the approximate pH of each of the mixtures formed in part (b).

pH of MgO–water mixture _____

pH of P_4O_{10}–water mixture _____

(ii) Give the formula of a compound that might be formed if the two mixtures formed in part (b) were to be combined.

Formula of compound _____ 3 marks

Total Marks: 11

6 **(a)** Deduce the oxidation state and co-ordination number of chromium in the complex compound $[Cr(H_2O)_4F_2]F$.

Oxidation state of chromium _____

Co-ordination number of chromium _____ 2 marks

(b) When the complex compound $[Cr(H_2O)_4F_2]F$ is treated with an excess of aqueous sodium hydroxide and hydrogen peroxide is added, a yellow solution is eventually formed. Treatment of this yellow solution with an excess of dilute sulfuric acid produces an orange solution.

Deduce the formulae of the chromium species present in the yellow solution and the formulae of the chromium species present in the orange solution, and write an equation for the conversion of the yellow species into the orange species.

Chromium species present in yellow solution _____

Chromium species present in orange solution _____

Equation _____ 3 marks

(c) A 0.234 g sample of a steel, containing iron and carbon only, was dissolved in dilute hydrochloric acid. The resulting solution was titrated with a 0.0185 mol dm^{-3} solution of potassium dichromate(VI) in the presence of an indicator. Exactly 35.8 cm^3 were required to reach the end-point.

(i) Explain why an indicator was necessary in this titration.

_____

_____

(ii) Construct the equation for the reaction between iron(II) ions and dichromate(VI) ions in acidic solution.

_____

_____

_____

(iii) Calculate the percentage of iron in the sample.

_____

_____

_____

_____

_____

_____

_____

_____

_____

_____

_____ 8 marks

Total Marks: 13

7 **(a)** Titanium(IV) chloride is a Lewis acid. It is a colourless liquid which reacts with moisture in air to form a white smoke.

    (i) What is meant by the term *Lewis acid*?

    _____

    _____

    (ii) Suggest why titanium(IV) chloride reacts with moisture in the air.

    _____

    _____ 3 marks

(b) (i) Write the outer electronic configurations of titanium(IV) and of titanium(III).

    Ti(IV) [Ar] _____

    Ti(III) [Ar] _____

    (ii) Titanium(IV) chloride is colourless, but titanium(III) chloride is a purple solid. Suggest why titanium(III) chloride is coloured and explain the origin of the colour.

    Reason for colour _____

    Origin of colour _____

    _____

    _____

    (iii) Suggest an instrumental method for determining the concentration of a solution of titanium(III) chloride.

    _____

    _____ 5 marks

(c) Titanium(III) ions are oxidised to titanium(IV) ions by potassium manganate(VII) in acidic solution. Writing titanium(III) ions as Ti^{3+} and titanium(IV) ions as Ti^{4+}, construct an ionic equation for this reaction.

_____

_____

_____ 2 marks

(d) Predict what you would see if an aqueous solution of titanium(III) chloride was treated with aqueous sodium carbonate. Write an equation for the overall reaction occurring.

Observations _____

_____

Equation _____

_____

_____ 4 marks

Total Marks: 14

8 (a) (i) When anhydrous copper(II) sulfate is added to water, a blue solution is formed. Write an equation for this reaction, showing the formula of the species responsible for the blue colour.

Equation _____

(ii) The addition of aqueous ammonia to the blue solution gives a blue precipitate, which dissolves to form a blue–violet solution when an excess of aqueous ammonia is added. Give the formula of the blue precipitate and the formula of the copper-containing species in the blue–violet solution.

Formula of blue precipitate _____

Formula of the copper species in the blue-violet solution _____

(iii) If ethylenediamine, $H_2NCH_2CH_2NH_2$, is added to this blue-violet solution, a further reaction occurs. Deduce the formula of the new species formed and explain why the reaction occurs.

Formula of new species _____

Reason reaction occurs _____

_____

_____ 6 marks

(b) The ammonia complex cisplatin, $Pt(NH_3)_2Cl_2$, has found a use in medicine. State its use and give a risk associated with its use.

Use of cisplatin _____

Risk of its use _____ 2 marks

(c) A solution of a silver compound in aqueous ammonia finds a use as a reagent in a laboratory test in organic chemistry.

(i) Give the formula of the silver species present in this solution.

_____

(ii) State the organic functional group which reacts in the test.

_____

(iii) Write an equation to show the reaction occurring in the test. Use the symbol R as the non-functional part of the organic molecule in the equation.

_____

_____ 4 marks

Total Marks: 12

9 **(a)** What is meant by the term *heterogeneous catalyst*?

_____

_____ 2 marks

(b) State the catalysts used in each of the following conversions:

(i) $2SO_2 + O_2 \rightarrow 2SO_3$

_____

(ii) $N_2 + 3H_2 \rightarrow 2NH_3$

_____

(iii) $CO + 2H_2 \rightarrow CH_3OH$

_____ 3 marks

(c) Give two reasons why the rhodium used in catalytic converters for cars is dispersed on a ceramic support.

Reason 1 _____

Reason 2 _____ 2 marks

(d) The aqueous reaction below, between iodide and persulfate ions,

$$2I^- + S_2O_8^{2-} \rightarrow I_2 + 2SO_4^{2-}$$

is slow in the absence of a catalyst.

(i) Explain why the reaction is slow _____

_____

(ii) Write equations to show how the reaction can be catalysed by Fe^{3+} ions.

_____

_____

_____ 3 marks

Total Marks: 10

10 (a) By stating the reagents required, show how each of the following conversions could be carried out.

(i) CrO_4^{2-} into $[Cr(OH)_6]^{3-}$

_____

_____

_____

_____

(ii) MnO_4^- into $[Mn(H_2O)_6]^{2+}$

(iii) $Cr_2O_7^{2-}$ into $[Cr(H_2O)_6]^{2+}$

_____ 8 marks

(b) A colour change occurs when aqueous cobalt(II) chloride is treated with an excess of concentrated hydrochloric acid.

(i) State the colour change and write an equation for the reaction occurring.

Colour change _____

Equation _____

(ii) Explain why the original colour of the aqueous cobalt(II) chloride is regenerated if an excess of water is then added to the mixture.

(iii) Suggest why there is no colour change when an excess of saturated aqueous sodium chloride is added to aqueous cobalt(II) chloride.

_____ 7 marks

_____ Total Marks: 15

Answers, explanations, hints and tips

Question	Answer		Marks
1 (a)	$\Delta H_1 \quad = +\Delta H_f$	(1)	
	elements $\quad Na(s) + \frac{1}{2} Br_2(l)$	(1)	
	$\Delta H_2 \quad = \frac{1}{2}\Delta H_{vap} + \frac{1}{2}\Delta H_{diss} + \Delta H_{sub}$	(1)	
	atoms $\quad Na(g) + Br(g)$	(1)	
	$\Delta H_3 \quad = \Delta H_i$	(1)	
	ion(s)1 $\quad Na^+(g) + Br(g) + e^-$	(1)	
	$\Delta H_4 \quad = \Delta H_{ea}$	(1)	
	ion(s)2 $\quad Na^+(g) + Br^-(g)$	(1)	
	$\Delta H_5 \quad = -\Delta H_L$	(1)	9
1 (b) (i)	$\Sigma\Delta H \text{ (cycle)} = 0$	(1)	
	$\therefore \frac{1}{2}\Delta H_{vap} + \frac{1}{2}\Delta H_{diss} + \Delta H_{sub} + \Delta H_i + \Delta H_{ea} + (-\Delta H_L) + (-\Delta H_f) = 0$		
	so $\quad \Delta H_{vap} = 2(-\frac{1}{2}\Delta H_{diss} - \Delta H_{sub} - \Delta H_i - \Delta H_{ea} + \Delta H_L + \Delta H_f)$	(1)	
	$= 2(-97 - 107 - 498 - (-325) + 753 - 361)$	(1)	
	$= +30 \text{ kJ mol}^{-1}$	(1)	
1 (b) (ii)	ionic	(1)	
	close agreement between Born–Haber and ionic model	(1)	
	no evidence for covalent bonding	(1)	7
1 (c) (i)	enthalpy change when 1 mol of gaseous ions	(1)	
	forms 1 mol of aqueous ions	(1)	
	under standard conditions	(1)	
1 (c) (ii)	first two equations reversed, third as in table *or* cycle	(1)	
	$\therefore K^+(g) \to K^+(aq) \quad \Delta H^{\ominus}_{hyd} = +364 - 718 + 17 \text{ kJ mol}^{-1}$	(1)	
	$\Delta H^{\ominus}_{hyd} = -337 \text{ kJ mol}^{-1}$	(1)	6
			Total 22
2 (a)	$\Delta G = \Delta H - T\Delta S$	(1)	1
2 (b)	$\Delta G \le 0$	(1)	1
2 (c) (i)	enthalpy of fusion evolved so ΔH –ve	(1)	
	more ordered product so ΔS –ve	(1)	
2 (c) (ii)	$-T\Delta S$ is +ve and outweighs ΔH for all T above 0 °C	(1)	
	ΔG for freezing > 0 if T above 0 °C	(1)	4
2 (d)	temperature drops, so enthalpy change ΔH is +ve	(1)	
	gas formed from solid, so ΔS is +ve	(1)	
	$-T\Delta S$ is –ve and will outweigh ΔH if T is big enough	(1)	
	room temperature high enough	(1)	4
2 (e) (i)	$\Delta S = \Sigma S \text{ (products)} - \Sigma S \text{ (reactants)}$	(1)	
	$= (3 \times 134 + 83) - (4 \times 112)$	(1)	
	$= +37.0 \text{ J K}^{-1} \text{ mol}^{-1}$	(1)	
2 (e) (ii)	$\Delta G = \Delta H - T\Delta S$		
	$= 16.8 - \dfrac{298 \times 37}{1000}$	(1)	
	$= 16.8 - 11.0 = +5.8$	(1)	
	kJ mol^{-1}	(1)	

Question	Answer		Marks
2 (e) (iii)	$T = \dfrac{\Delta H}{\Delta S}$	(1)	
	$= \dfrac{16.8 \times 10^3}{37.0}$	(1)	
	$= 454$ K	(1)	9
			Total 19
3 (a)	electron donor	(1)	1
3 (b) (i)	standard hydrogen electrode	(1)	
	0 V	(1)	
3 (b) (ii)	solutions: 1 mol dm^{-3}	(1)	
	temperature: 298 K	(1)	
	pressure: 100 kPa	(1)	
	measured: with zero current flowing	(1)	
3 (b) (iii)	to allow transfer of ions and to create an electrical circuit	(1)	
	any soluble unreactive salt such as KCl or KNO_3	(1)	8
3 (c)	list of standard reduction potentials	(1)	
	in (numerical) order	(1)	2
3 (d) (i)	$Zn^{2+}(aq) + 2e^- \rightarrow Zn(s)$	(1)	
3 (d) (ii)	$Zn(s) \,|\, Zn^{2+}(aq) \,\|\, Fe^{2+}(aq) \,|\, Fe(s)$		
	species (with state symbols) in correct order	(1)	
	phase boundaries and salt bridge correct	(1)	
	e.m.f. $-0.44 - (-0.76) = +0.32$ V	(1)	
3 (d) (iii)	Fe^{2+}	(1)	
	Fe^{2+}/Fe couple has the more positive potential	(1)	
	so gains electrons more easily	(1)	7
			Total 18
4 (a) (i)	not rechargeable	(1)	
4 (a) (ii)	cheap	(1)	
	disposal hazard to environment	(1)	
4 (a) (iii)	$Li(s) \rightarrow Li^+ + e^-$	(1)	
	$MnO_2(s) + e^- \rightarrow MnO_2^-(s)$	(1)	5
4 (b)	*increase* $[NH_4^+(aq)]$ and/or $[OH^-(aq)]$	(1)	
	decrease $[Zn(NH_3)_4]^{2+}(aq)]$ *(decrease [H$_2$O(l)] loses mark)*	(1)	
	decrease temperature	(1)	3
4 (c) (i)	rechargeable cell	(1)	
	reaction is reversed on charging	(1)	
4 (c) (ii)	two equations rearranged	(1)	
	$PbO_2(s) + 3H^+(aq) + HSO_4^-(aq) + 2e^- \rightarrow PbSO_4(s) + 2H_2O(l)$ $+ 1.68$ V		
	$PbO_2(s) + 2H^+(aq) + 2HSO_4^-(aq) + Pb(s) \rightleftharpoons 2PbSO_4(s) + 2H_2O(l)$ $+ 2.04$ V		
	subtract:	(1)	
	$PbSO_4(s) + H^+(aq) + 2e^- \rightarrow Pb(s) + HSO_4^-(aq)$	(1)	
	$E = 1.68 - 2.04 = -0.36$ V	(1)	6
4 (d) (i)	$2H_2O(l) + 2e^- \rightarrow H_2(g) + 2OH^-(aq)$	(1)	
4 (d) (ii)	$O_2(g) + 2H_2O(l) + 4e^- \rightarrow 4OH^-(aq)$	(1)	

Question	Answer		Marks
4 (d) (iii)	two equations rearranged	(1)	
	$O_2(g) + 2H_2O(l) \rightarrow 4OH^-(aq) - 4e^-$ $E^{\ominus} = +0.40$ V		
	$2H_2(g) + 4OH^-(aq) \rightarrow 4H_2O(l) + 4e^-$ $E^{\ominus} = +0.83$ V		
	add:	(1)	
	$2H_2(g) + O_2(g) \rightarrow 2H_2O(l)$ *cell e.m.f.* = +1.23 V	(1)	
4 (d) (iv)	3 mol gas → 2 mol liquid	(1)	
	high pressure favours forward reaction	(1)	
	exothermic	(1)	
	but low T makes reaction too slow	(1)	9
			Total 23
5 (a)	ionic lattice	(1)	
	molecular	(1)	
	ionic lattices have strong electrostatic forces between ions	(1)	
	molecular solids have weak forces between molecules	(1)	4
5 (b)	nothing or slight solubility	(1)	
	$MgO + H_2O \rightarrow Mg^{2+} + 2OH^-$	(1)	
	violent reaction or smoke, dissolves	(1)	
	$P_4O_{10} + 6H_2O \rightarrow 4H_3PO_4$	(1)	4
5 (c) (i)	8 – 10	(1)	
	0	(1)	
5 (c) (ii)	$MgHPO_4$ *or* $Mg(H_2PO_4)_2$ *or* $Mg_3(PO_4)_2$	(1)	3
			Total 11
6 (a)	+3	(1)	
	6	(1)	2
6 (b)	CrO_4^{2-}	(1)	
	$Cr_2O_7^{2-}$	(1)	
	$2CrO_4^{2-} + 2H^+ \rightarrow Cr_2O_7^{2-} + H_2O$	(1)	3
6 (c) (i)	no clear colour change at end-point	(1)	
6 (c) (ii)	$Cr_2O_7^{2-} + 14H^+ + 6e \rightarrow 2Cr^{3+} + 7H_2O$		
	$6Fe^{2+} \rightarrow 6Fe^{3+} + 6e^-$		
	both half-equations (*1*) (*if complete equation not given below*)		
	$Cr_2O_7^{2-} + 6Fe^{2+} + 14H^+ \rightarrow 2Cr^{3+} + 7H_2O + 6Fe^{3+}$	(2)	
6 (c) (iii)	moles $Cr_2O_7^{2-} = 35.8 \times 0.0185 \times 10^{-3} = 6.62 \times 10^{-4}$	(1)	
	moles $Fe^{2+} = 6 \times 6.62 \times 10^{-4}$	(1)	
	grams Fe $= 55.8 \times 6 \times 6.07 \times 10^{-4}$	(1)	
	$= 0.221$		
	% Fe $= \dfrac{0.221 \times 100}{0.234}$	(1)	
	$= 94.6\%$	(1)	8
			Total 13
7 (a) (i)	electron-pair acceptor	(1)	
7 (a) (ii)	water has lone-pair electrons	(1)	
	it can donate these to a Lewis acid *or* is a Lewis base	(1)	3
7 (b) (i)	$3d^0$	(1)	
	$3d^1$	(1)	

Question	Answer		Marks
7 (b) (ii)	titanium(III) chloride has a d electron	(1)	
	electron can be promoted to a higher energy level by absorption of visible light	(1)	
7 (b) (iii)	visible spectroscopy *or* colorimeter	(1)	5
7 (c)	$MnO_4^- + 8H^+ + 5e \rightarrow Mn^{2+} + 4H_2O$		
	$\qquad 5Ti^{3+} \rightarrow 5Ti^{4+} + 5e^-$		
	both half equations (1) (*if complete equation not given below*)		
	$MnO_4^- + 8H^+ + 5Ti^{3+} \rightarrow 5Ti^{4+} + Mn^{2+} + 4H_2O$	(2)	2
7 (d)	precipitate + effervescence	(1)	
	$[Ti(H_2O)_6]^{3+} + 3H_2O \rightarrow [Ti(OH)_3(H_2O)_3] + 3H_3O^+$	(1)	
	$2H_3O^+ + CO_3^{2-} \rightarrow CO_2 + 3H_2O$	(1)	
	$2[Ti(H_2O)_6]^{3+} + 3CO_3^{2-} \rightarrow 2[Ti(OH)_3(H_2O)_3] + 3CO_2 + 3H_2O$	(1)	4
			Total 14
8 (a) (i)	$CuSO_4(s) + 6H_2O(l) \rightarrow [Cu(H_2O)_6]^{2+}(aq) + SO_4^{2-}(aq)$	(1)	
8 (a) (ii)	$[Cu(OH)_2(H_2O)_4]$ *or* $Cu(OH)_2$	(1)	
	$[Cu(NH_3)_4(H_2O)_2]^{2+}$	(1)	
8 (a) (iii)	$[Cu(H_2NCH_2CH_2NH_2)_2(H_2O)_2]^{2+}$	(1)	
	ethylenediamine is a bidentate ligand	(1)	
	$4NH_3$ liberated but only $2H_2NCH_2CH_2NH_2$ used so increase in entropy *or*		
	chelate effect	(1)	6
8 (b)	cancer cure *or* chemotherapy	(1)	
	side effects severe *or* hair loss	(1)	2
8 (c) (i)	$[Ag(NH_3)_2]^+$	(1)	
8 (c) (ii)	$-CHO$	(1)	
8 (c) (iii)	$RCHO + 2[Ag(NH_3)_2]^+ + 3OH^- \rightarrow RCOO^- + 2Ag + 4NH_3 + 2H_2O$	(2)	4
			Total 12
9 (a)	increases rate of attainment of equilibrium	(1)	
	in different phase from reactants	(1)	2
9 (b) (i)	V_2O_5	(1)	
9 (b) (ii)	Fe	(1)	
9 (b) (iii)	Cr_2O_3	(1)	3
9 (c)	increases surface area	(1)	
	more economical	(1)	2
9 (d) (i)	two negative ions repel each other *or* do not wish to collide	(1)	
9 (d) (ii)	$2Fe^{3+} + 2I^- \rightarrow I_2 + 2Fe^{2+}$	(1)	
	$2Fe^{2+} + S_2O_8^{2-} \rightarrow 2Fe^{3+} + 2SO_4^{2-}$	(1)	3
			Total 10
10 (a) (i)	add dilute HCl *or* H_2SO_4 and $FeSO_4$ (*or* SO_2)	(2)	
	excess Na(OH)	(2)	
10 (a) (ii)	add dilute HCl *or* H_2SO_4 and $FeSO_4$ (*or* SO_2 *or* Zn)	(2)	
10 (a) (iii)	add Zn and dilute HCl *or* H_2SO_4 (keeping air out)	(2)	8
10 (b) (i)	pink to blue	(2)	
	$[CoH_2O)_6]^{2+} + 4Cl^- \rightarrow [CoCl_4]^{2-} + 6H_2O$	(2)	
10 (b) (ii)	equilibrium reaction, driven to left by excess H_2O	(1)	
10 (b) (iii)	insufficient Cl^- *or* NaCl not sufficiently soluble	(1)	
	to drive equilibrium to right	(1)	7
			Total 15

The table below highlights aspects of *How Science Works* in the exemplar questions.

Question	How Science Works
1 (b)/(c)	analysis and interpretation of experimental data
2 (d)	use of theory to explain observations
2 (e)	use of experimental data to make a prediction
3 (d)	communicate information using appropriate terminology
4 (a)	consider benefits and risks of disposable cells
6	analysis and interpretation of experimental data
7 (b) (ii)	suggest a suitable instrumental method of analysis
7 (d)	use knowledge to make a prediction
8 (b)	consider the benefits and risks of *cisplatin*
9 (c)	understand the use of catalysts and the economics behind their use
10 (a)	select synthetic reagents
10 (b)	explanation of experimental observations

Glossary

$\Delta H^{\ominus}$ (298 K)	the standard enthalpy change at 298 K
absolute entropy	see *standard entropy*
acid dissociation constant (K_a)	the equilibrium constant for the dissociation of a weak acid in water; sometimes called the *acidity constant*
acidity constant	see *acid dissociation constant*
acidity reaction	a reaction of a metal aqua-ion in which an O—H bond in a co-ordinated water molecule is broken, releasing H^+; also known as the *hydrolysis reaction*
adsorption	the process by which a substance (a liquid or a gas) is weakly bonded to and held in place on a solid surface
affinity	see *electron affinity*
amphoteric character	the ability to react with both acids and bases
amphoterism	the property of being able to react with both acids and bases
aqua ion	a metal ion surrounded by water ligands
autocatalysis	a process in which a reaction is catalysed by one of its products
Beer–Lambert law	$A = \varepsilon cl$ where absorbance (A), concentration (c), cell path-length (l) and molar absorption coefficient (ε) are linked
bidentate	able to donate a pair of electrons from each of two separate atoms
bond dissociation enthalpy ($\Delta H^{\ominus}_{diss}$)	the standard molar enthalpy change for the breaking of a covalent bond in a gaseous molecule to form two gaseous free radicals, e.g. for the process $Cl_2(g) \rightarrow 2Cl\bullet(g)$ *or* $CH_4(g) \rightarrow \bullet CH_3(g) + H\bullet(g)$
carbon monoxide (CO)	a lethal, colourless, odourless gas that results from the incomplete combustion of carbon and its compounds
carboxyhaemoglobin	a very stable complex of haemoglobin; the affinity of haemoglobin for carbon monoxide is much greater than its affinity for oxygen, so that this complex is formed in preference to *oxyhaemoglobin*, thereby inhibiting the uptake of oxygen by red blood cells and making carbon monoxide a very dangerous gas
catalyst	a substance that alters the rate of a reaction without itself being consumed
catalyst poison	an unwanted contaminant or waste product which is adsorbed too strongly on to a catalyst surface, thereby preventing it from acting efficiently
catalytic adsorption	weakly-bonded molecules are held on a solid catalyst surface in a manner that decreases the activation energy for reaction or increases the encounter rate (collision rate) between reacting species
catalytic converters	used in cars; contain catalysts capable of converting harmful gaseous products into less harmful ones
cell convention	the cell diagram is written with the more positive electrode (the one at which reduction occurs) shown as the right-hand electrode
cell diagram	has two electrodes back to back, joined by a salt bridge that is conventionally denoted by two vertical bars, e.g. for a zinc/copper cell, the diagram is $Zn(s) \mid Zn^{2+}(aq) \parallel Cu^{2+}(aq) \mid Cu(s)$

cell e.m.f	is given by *cell e.m.f.* $= E^{\ominus}(R) - E^{\ominus}(L)$
cell potential	see *standard e.m.f.*
cell reaction	with the more positive electrode shown as the right-hand electrode, the cell reaction goes in the forward direction
change of colour in a complex	can arise from a change in oxidation state, from a change in co-ordination number or from a change in ligand
charge-to-size ratio	the ratio of the overall charge on an ion to its size
chelate effect	the effect of driving a reaction towards products in the reaction when a bidentate or a multidentate ligand reacts with a co-ordination compound surrounded by unidentate ligands (see *entropy-driven reaction*)
cisplatin	an anti-cancer drug with the formula $[Pt(NH_3)_2Cl_2]$
colorimeter	an instrument for measuring the intensity of colours
combined heat and power (CHP) systems	domestic electricity generation systems which use up waste heat to provide local heating; these systems are made more cost-effective by feeding excess electrical power, during 'idle' periods, into the electricity grid, thus running meters 'backwards'
complex	see *co-ordination compound*
Contact process	the industrial process for converting a mixture of sulfur dioxide and oxygen into sulfur trioxide; uses a vanadium(V) oxide catalyst
co-ordinate bond	a covalent bond formed when the pair of electrons originate from one atom; a straight arrow is drawn to show the origin of the electron pair in the co-ordinate bond, e.g. as in $F_3B \leftarrow NH_3$
co-ordination compound	a compound formed when one or more ligands bond to a metal ion, commonly called a *complex*
co-ordination number	the number of atoms bonded to a metal ion
couple	see *redox couple*
d-block elements	elements in the Periodic Table that have their highest energy electrons in a d sub-level
direction of a cell reaction	a cell reaction goes forwards, in the direction written, if the corresponding *cell e.m.f.* is positive
disposable cell	see *non-rechargeable cell*
disproportionation	the term applied to a reaction in which the same species is simultaneously oxidised and reduced
EDTA^{4-}	the bis[di(carboxymethyl)amino]ethane ion, a *hexadentate* ligand with many uses in analytical and industrial chemistry; forms very stable complexes
electrochemical cell	contains two electrodes immersed in an electrolyte
electrochemical series	a list of standard electrode potentials, $E^{\ominus}$, arranged in order of their numerical values
electrode compartment	an electrode immersed in its associated electrolyte
electrode reaction	the half-reaction, oxidation or reduction, that occurs in a given cell compartment, e.g. for the process $Zn(s) \rightarrow Zn^{2+}(aq) + 2e^-$ *or* $Cu^{2+}(aq) + 2e^- \rightarrow Cu(s)$
electrode representation	the notation used to represent an electrode in a cell diagram, e.g. the standard Cu/Cu^{2+} couple is written: $Cu(s) \mid Cu^{2+}(aq, 1.00 \text{ mol dm}^{-3})$ $\qquad E^{\ominus} = +0.34$ V at 298 K

electron affinity ($\Delta H_{ea}^{\ominus}$)	the standard molar enthalpy change for the addition of an electron to an isolated atom in the gas phase, e.g. for the process $Cl(g) + e^-(g) \rightarrow Cl^-(g)$
enthalpy of atomisation ($\Delta H_{at}^{\ominus}$)	the standard enthalpy change for the formation of one mole of gaseous atoms from an atomic element in its standard state, e.g. for the process $Na(s) \rightarrow Na(g)$
enthalpy change (ΔH)	the amount of heat released or absorbed when a chemical or physical change occurs at constant pressure (see also *standard enthalpy change*)
enthalpy of formation	see *standard enthalpy of formation*
enthalpy of fusion ($\Delta H_{fus}^{\ominus}$)	the standard molar enthalpy change when a solid forms a liquid at its melting point, e.g. for the process $H_2O(s) \rightarrow H_2O(l)$
enthalpy of hydration ($\Delta H_{hyd}^{\ominus}$)	the standard molar enthalpy change for the formation of aqueous ions from gaseous ions, e.g. for the process $Mg^{2+}(g) \xrightarrow{\text{water}} Mg^{2+}(aq)$
enthalpy of lattice dissociation ($\Delta H_L^{\ominus}$)	the standard molar enthalpy change for the separation of a solid ionic lattice into its gaseous ions, e.g. for the process $NaCl(s) \rightarrow Na+(g) + Cl^-(g)$, which is *endothermic*
enthalpy of lattice formation ($\Delta H_L^{\ominus}$)	the standard molar enthalpy change for the formation of a solid ionic lattice from its gaseous ions, e.g. for the process $Na^+(g) + Cl^-(g) \rightarrow NaCl(s)$, which is *exothermic*
enthalpy of solution ($\Delta H_{sol}^{\ominus}$)	the standard molar enthalpy change that occurs when an ionic solid dissolves in enough water to ensure that the dissolved ions are well separated and do not interact with one another, e.g. for the process $NaCl(s) \xrightarrow{\text{water}} Na^+(aq) + Cl^-(aq)$
enthalpy of sublimation ($\Delta H_{sub}^{\ominus}$)	the standard molar enthalpy change that occurs on sublimation, when a solid changes directly to a gas without forming a liquid phase
enthalpy of vaporisation ($\Delta H_{vap}^{\ominus}$)	the standard molar enthalpy change when a liquid forms a gas at its boiling point, e.g. for the process $H_2O(l) \rightarrow H_2O(g)$
entropy	a measure of the disorder in a system
entropy-driven reaction	one where a large positive entropy change dominates a much smaller enthalpy change, making the value of ΔG for the reaction large and negative, resulting in a very favourable (feasible) process, e.g. reactions of $EDTA^{4-}$
excited state	an energy state higher than the ground state
feasibility temperature	the equilibrium temperature, T, at which $\Delta G = 0$: $T = \dfrac{\Delta H^{\ominus}}{\Delta S^{\ominus}}$
feasible (spontaneous) change	one that has a natural tendency to occur without being driven by any external influences; in terms of Gibbs free-energy, $\Delta G \leq 0$ for feasible change
first ionisation enthalpy (energy) ($\Delta H_i^{\ominus}$)	the standard molar enthalpy change for the removal of an electron from an atom in the gas phase to form a positive ion and an electron, both also in the gas phase, e.g. for the process $Na(g) \rightarrow Na^+(g) + e^-(g)$
free radical	a species that results from the homolytic fission of a covalent bond; it contains an unpaired electron

fuel cell	one in which produces electrical power from an external supply of a fuel and an oxidant
gas electrode	an inert metal (usually platinum) surrounded by a gas in equilibrium with a solution of its ions (see *standard hydrogen electrode*)
Gibbs free-energy change ($\Delta G^{\ominus}$)	Determines the direction of spontaneous change by combining the influences of enthalpy and of entropy, through the relationship $\Delta G^{\ominus} = \Delta H^{\ominus} - T\Delta S^{\ominus}$
Gibbs free-energy change at equilibrium	for a system at equilibrium, $\Delta G = 0$
ground state	the lowest energy state of electrons in a species, e.g. of d electrons in a transition-metal complex
Haber process	the industrial process for converting a mixture of nitrogen and hydrogen into ammonia; uses an iron catalyst
haemoglobin	an octahedrally co-ordinated iron(II) complex, responsible for the red colour of blood and for the transport of oxygen by red blood cells from one part of the body to another
half-equation	a balanced equation for an oxidation or a reduction that shows a species gaining or losing electrons, e.g. for the process $MnO_4^- + 8H^+ + 5e^- \longrightarrow Mn^{2+} + 4H_2O$
heterogeneous catalysis	proceeds through the adsorption of reactants on to a catalytic surface
heterogeneous catalyst	acts in a different phase from the reactants
heterolytic fission	formation of ions when a covalent bond breaks with an unequal splitting of the bonding pair of electrons, e.g. for the process $HCl \longrightarrow H^+ + Cl^-$
hexadentate ligand	having six electron pairs capable of forming co-ordinate bonds (see *EDTA^{4-}*)
homogeneous catalysis	proceeds through the formation of an intermediate species in the same phase as the reactants
homogeneous catalyst	acts in the same phase as the reactants
homolytic fission	formation of radicals when a covalent bond breaks with an equal splitting of the bonding pair of electrons, e.g. for the process $HCl \longrightarrow H\bullet + Cl\bullet$
hydrogen electrode	corresponds to the redox couple H^+/H_2 at a platinum surface
hydrolysis reaction	see *acidity reaction*
inert support medium	an unreactive solid used to dilute a reagent or a catalyst
irreversible cell	see *non-rechargeable cell*
IUPAC convention for electrode reaction half-equations	all redox half-equations are written as *reductions* (electron gain)
lattice enthalpy	see *enthalpy of lattice dissociation* and *enthalpy of lattice formation*
Lewis acid	an electron-pair acceptor; metal cations can act as Lewis acids
Lewis base	an electron-pair donor; ligands and nucleophiles can act as Lewis bases

ligand	an atom, ion or molecule capable of donating one or more pairs of electrons to a metal ion; a ligand acts also as a Lewis base or a nucleophile
ligand substitution	replacement of one or more ligands by others
linear complex	a metal ion with two ligands in a straight line on either side; the two bond angles are each 180°
mean bond enthalpy (ΔH_B)	the average of several values of the bond dissociation enthalpy for a given type of bond, taken from a range of different compounds.
metal electrode	conventionally written as $M^{n+}(aq)/M(s)$, or the reverse
mixed catalysts	those prepared by mixing two or more catalytic substances
multidentate ligand	able to donate a pair of electrons from each of several separate atoms
non-rechargeable cell	one not intended to be recharged by an electric current; also known as a *primary cell* or as a *disposable cell*
nucleophile	an electron-rich molecule or ion able to donate a pair of electrons (see *Lewis base* and *ligand*)
octahedral complex	a central metal ion with six ligands lying at the corners of a octahedron; the six bond angles in a regular octahedron are each 90°
overall equation for a redox reaction	obtained by adding together two half-equations and balancing the numbers of electrons gained and lost in each half-equation
oxidation	the process of electron loss
oxidation state	the charge a central atom in a complex would have if it existed as a solitary simple ion without bonds to other species
Period 3 elements	elements in the third period (row) of the Periodic Table
pK_a	logarithmic expression of the acid dissociation constant in aqueous solution $pK_a = -\log_{10} K_a$
Planck's constant (h)	the constant of proportionality between the energy (ΔE) of absorbed light and its frequency (v) in the relationship $\Delta E = hv$, where h has the value 6.63×10^{-34} J s
polarising power	the ability of a metal ion to distort the electron cloud of a neighbouring anion
primary cell	see *non-rechargeable cell*
rechargeable cell	one designed to be recharged by an electric current; also known as a *secondary cell*
redox couple	a shorthand way of writing a reduction half-equation, e.g. for the process $Cu^{2+}(aq) + 2e^- \rightarrow Cu(s)$, the redox couple is Cu^{2+}/Cu
redox electrode	one at which two oxidation states of a given element undergo a reduction reaction at an inert metal surface, e.g. for the Fe^{3+}/Fe^{2+} couple the electrode is written $Pt(s) \mid Fe^{2+}(aq), Fe^{3+}(aq)$
redox potential	the standard electrode potential for a reduction process, e.g. $Cu^{2+}(aq) + 2e^- \rightarrow Cu(s)$
redox titration	used in volumetric analysis to determine the concentration of either an oxidising agent or a reducing agent, e.g. iron(II) with manganate(VII) ions
reduction	the process of electron gain
reference temperature	under standard conditions, this temperature is most commonly taken to be 298 K

salt bridge	an electrolyte solution used to complete electrical contact between two electrode compartments; it allows the transfer of ions between compartments
second ionisation enthalpy	the standard molar enthalpy change for the removal of an electron from a singly positively charged ion in the gas phase to form a gaseous di-positive ion and an electron, both also in the gas phase, e.g. for the process $Na^+(g) \rightarrow Na^{2+}(g) + e^-(g)$
secondary cell	see *rechargeable cell*
sequestering ability	(*to sequester* – literally *to give up for safekeeping*) – the property of a substance to keep metal ions in solution even when anions which normally cause precipitation (such as OH^- or $CO_3{}^{2-}$) are added
spontaneous (feasible) change	one that has a natural tendency to occur without being driven by any external influences; in terms of Gibbs free-energy, $\Delta G \leq 0$ for spontaneous change
square-planar complex	a central metal ion with four ligands lying in one plane at the corners of a square; the four bond angles in such a square are each 90°
standard amount	the mole
standard cell conditions	100 kPa, 298 K, 1.00 mol dm^{-3}
standard changes	changes measured under standard conditions and signified by the use of a superscript *plimsoll sign*, $\ominus$
standard conditions	refer to a standard pressure of 100 kPa at a stated reference temperature (most commonly 298 K); for solutions, also at a concentration of 1.00 mol dm^{-3}
standard electrode potential ($E^\ominus$)	the standard potential of a cell measured with the standard hydrogen electrode as the left-hand electrode and the unknown as the right-hand one
standard e.m.f.	the potential difference between the electrodes of a standard electrochemical cell measured under zero-current conditions
standard enthalpy change ($\Delta H^\ominus$)	the change in enthalpy when reactants in their standard states form products also in their standard states
standard enthalpy of formation ($\Delta H_f^\ominus$)	the enthalpy change under standard conditions when one mole of a compound is formed from its elements with all reactants and products in their standard states
standard entropy ($S^\ominus$)	an absolute measure of entropy, based on zero entropy at zero Kelvin
standard hydrogen electrode	has a defined potential of 0 V and an electrode representation: $Pt(s) \mid H_2(g, 1 \text{ bar}) \mid H^+(aq, 1.00 \text{ mol } dm^{-3})$ \qquad $E = 0$ at 298 K
standard state	the pure, most stable, form of a substance at a given temperature and 100 kPa
sublimation	the process by which a substance changes directly from a solid to a gas without forming a liquid phase, e.g. for the process $I_2(s) \rightarrow I_2(g)$
substitution reaction	one in which one or more ions or molecules are replaced by other ions or molecules, e.g. the replacement of one ligand by another in a reaction of a transition-metal complex
surface-to-mass ratio	the ratio of the area of a surface to the mass of the substance covering it
tetrahedral complex	a central ion with four ligands lying at the corners of a tetrahedron; the four bond angles in a regular tetrahedron are each 109.5°

titrations using potassium dichromate(VI)	when used as an oxidising agent in acidic solution, there is no distinctive colour change at the end-point, so that an indicator is needed; commonly, sodium diphenylaminesulfonate is used, which turns from colourless to purple
titrations using potassium manganate(VII)	when used as an oxidising agent in acidic solution, no additional indicator is required; at the end-point, the presence of even a slight excess of the dark purple manganate(VII) ion is used to indicate complete reaction as the solution turns pink
transition element	an element having an incomplete d (or f) sub-level either in the element or in one of its common ions
transition-metal complex redox reaction	one in which a transition-metal ion changes its oxidation state, either being oxidised or being reduced
unidentate	able to donate a pair of electrons from one atom only
visible spectrophotometer	a device that uses visible light of varying frequencies to measure the amount of light absorbed by a coloured solution; the absorbed amount is proportional to the concentration of the absorbing species in the solution under test

Notes